MANUFACTURING RESEARCH AND TECHNOLOGY 23

FLEXIBLE MANUFACTURING SYSTEMS: RECENT DEVELOPMENTS

MANUFACTURING RESEARCH AND TECHNOLOGY

MANUFACTURING RESEARCH AND TECHNOLOGY 23

Flexible Manufacturing Systems: Recent Developments

edited by

A. Raouf
M. Ben-Daya
Systems Engineering Department
King Fahd University of
*   Petroleum and Minerals*
Dhahran, Saudi Arabia

ELSEVIER
Amsterdam – Lausanne – New York – Oxford – Shannon – Tokyo 1995

ELSEVIER SCIENCE B.V.
Sara Burgerhartstraat 25
P.O. Box 211, 1000 AE Amsterdam, The Netherlands

Library of Congress Cataloging-in-Publication Data

Flexible manufacturing systems : recent developments / edited by A.
   Raouf and M. Ben-Daya.
        p.   cm. -- (Manufacturing research and technology ; 23)
    Includes bibliographical references.
    ISBN 0-444-89798-4
    1. Flexible manufacturing systems.   I. Raouf, A. (Abdul), 1929-
. II. Ben-Daya, M. (Mohamed)   III. Series.
TS155.65.F59   1995
670.42'7--dc20   94-48220
                                                                CIP

ISBN: 0 444 89798 4

This book is printed on acid-free paper.

Printed in The Netherlands

Preface

This volume contains new and updated material from *Flexible Manufacturing: Recent Developments in FMS, Robotics, CAD/CAM, CIM* which was a selection of papers presented at the VIIth International Conference on Production Research, held in Windsor, Ontario during 1983. This book was the first of the series titled "Manufacturing Research and Technology", published by Elsevier. The book was very well received by scholars and researchers interested in attending to problems related to Flexible Manufacturing Systems.

The present book comprises of five parts and these are : (1) FMS in Perspective, (2) Flexibility Issues, (3) FMS Planning, (4) FMS Control, and (5) FMS Applications.

In the first Part, there are two papers. In the first paper, a review of the pertinent literature with the objective of identifying the applications of existing FMS models and future research directions, is presented. The second paper gives an outline contents, main reasons, planning and developing procedures, success determining factors and results of 100 FMS related projects carried out by the authors.

The second Part contains the papers related to Flexibility Issues. The first paper describes some of the salient features affecting the flexibility of FMS. A methodology of quantifying overall flexibility is suggested. The second paper attempts to suggest methods for calculating different flexibility levels, strategies for a more flexible view upon products and processes and presents examples from Swedish industry. The third paper presents ways to improve the flexibility of a single-stage production-system having different characteristics.

The third Part focuses on the FMS Planning issues and contains five papers. The first paper presents a review of FMS short term planning problems. It contains related models and cites directions for future research. The second paper presents a planning and scheduling methodology for FMS. It presents four models and discusses some of the algorithms for solving the loading problems. The third paper considers the problem of selecting optimal routes for manufacturing various part types and outlines an hierarchical solution procedure using an example. The fourth paper deals with the heuristics of loading of FMS for a given set of part types chosen for immediate simultaneous production, allocates the operations and associated tooling for these part types among the machines subject to capacity and technological constraints. Algorithms that present reasonable solutions to the FMS loading problems are developed. The fifth paper presents a framework for developing maintenance policy for FMS. In addition, unique characteristics of FMS from maintenance management point of view are presented.

The fourth Part contains two papers dealing with FMS control. In the first paper, FMS planning & control is treated as a control problem. The analysis phase of the paper contains modelling tools needed to state the hierarchical planning system and the control System. Synthesis phase defines the methodology for specifying and developing software. The second paper presents a modern systems theory for the design of sequencing controllers when a controller is considered as a separate entity from the work cell.

The fifth Part contains two papers and presents application of FMS. The first paper describes the current status of development of an integrated robotic flexible welding cell. The second paper describes the practice of FMS at Toyota.

This book will be of much interest to researchers, managers and students of FMS.

We are grateful to the authors and the reviewers for assisting us in preparing this volume which was supported by the King Fahd University of Petroleum & Minerals, Dhahran. The editors acknowledge the assistance provided in this regard.

A. Raouf

M. Ben-Daya

Contents

Part I

FMS in Perspective

Flexible Manufacturing Systems: Recent Developments
A. Raouf and M. Ben-Daya (Editors)
1995 Elsevier Science B.V.

Flexible Manufacturing Systems: An Investigation For Research And Applications

A. Gunasekaran, T. Martikainen and P. Yli-Olli

School of Business Studies, University of Vaasa,
65101 Vaasa, Finland

Abstract

Flexible Manufacturing Systems(s) (FMS) have already proved their great success in a large number of manufacturing industries. Realising the importance of FMS in increasing productivity and quality, an attempt has been made in this paper to review the literature available on FMS with the objective to identify the applications of existing FMS models and future research directions in the areas of FMS.

1 Introduction

Excellence in manufacturing systems has been recognized as a major factor behind the success of industrial or manufacturing firms. New technologies of manufacturing processes play a significant role in this. Achieving the full potential of these new production technological innovations, however, necessitates a broad range of management, engineering, and systems issues (Stecke [177]). As a result, the implementation of modern manufacturing methods and technologies represents an opportunity for significant contributions from the fields of Operations Research (OR) and Management Science (MS). Further, the growth in demand for the products coupled with the intense competition in the market and the concern for the product's quality led the manufacturing industries to devise and implement new manufacturing technologies which include the development of FMS. The purpose of this paper is to (i) classify the literature available on FMS based on the nature and application of the models; (ii) review the literature available on FMS with an objective to identify the gap between theory and practice; and (iii) point out future research directions in the areas of FMS.

The organization of this paper is as follows: Section 2 deals with an introduction to the design and operational issues of FMS. The classification scheme proposed for the FMS literature is presented in Section 3. Section 4 reviews the previous research papers on FMS. The limitations of the existing FMS models and future research directions are presented in Section 5. The conclusion of this research are given in Section 6.

2 Flexible manufacturing systems

Traditionally, production facilities have two conflicting objectives: flexibility and productivity. Flexibility refers to producing a number of distinct products in a job shop environment where opportunities for production variability exist. Productivity, on the other hand, refers to high speed production that is similar to an assembly line. Studies have demonstrated that the productivity in a job shop environment is low as compared to flow shop (Hutchinson [84]). Therefore, increasing job shop productivity while maintaining production flexibility has been a desired goal of the industries. Emergency of FMS is a development in this direction.

An FMS can be characterized as a set of flexible machine tools connected by a material handling system and which is controlled by both computers and human operators (Buzacott and Yao [30]). Every material handling system has an automatic part-transportation system. Some advanced systems also contain automatic tool transportation devices. These can transfer tools among tool magazines and the central tool storage area while the system is in operation. This advancement in FMS hardware has rendered a major impact on FMS operation (Edghill and Davies [54], Jaikumar [89], [90]). Applications of these methods facilitate the ability to process many variations within a single product family and make speedy extensions of an existing product line ([115]).

In general, FMS operation decisions compose of pre-release decisions and post-release decisions. Pre-release decisions denote the prearrangement of parts and tools before the FMS begins to process. Post-release decisions, on the other hand, denote the sequencing and routing of parts when the system is in operation (Stecke [179]). The basic decision problems in FMS can be grouped into design problems (equipment section, layout, materials handling, etc.) and operational problems (aggregate planning, part selection, resource grouping, production ratio determination, resource allocation and loading). Details of these problems can be found in Stecke [179] and Kusiak [105]. The complexity of these problems depends on whether the FMS is of a dedicated type or a random type. Usually, a dedicated-type system is designed to produce a rather small family of similar parts with a known and limited variety of processing requirements, while in a random-type system a large family of parts having a wide range of characteristics is produced. In the random-type system, the product mix is not accurately defined at the time of installing the system.

3 Classification of FMS literature

During the last decade, a number of research papers has been published to deal with different aspects of FMS. A classification scheme based on the nature and application of the models is produced for an easy understanding of the research work on FMS. In the review of FMS modelling approaches, Kalkunte et al. [91] classified the modelling of an FMS in four levels:

1. strategic analysis and economic justification, which provides long-range, strategic business plans [4,12,26,46,63,81,129,130,132,140,142,168,173,174,185,186,203];

2. facility design, in which strategic business plans are coalesced into a specific facility design to accomplish long-term managerial objectives [2,6,9,10,14,16, 18,21,23,24,27]- [29,32]- [35,44,46,48,52,53,55]- [57,59,60,62]- [64,67,68,71,75,76, 83,84,89]- [91,94,95]- [97,102,106,108,110,111,113,114,117,118,120,122]- [125,130, 132,139]- [141,143,147,148,150,153,155,165,166,170,171,173,188,189,191]- [197,209, 216]

3. intermediate-range planning, which encompasses decisions related to master production scheduling and deals with a planning horizon from several days to several months in duration [1,7,10,11,13,15,20,25,36,37,39,40,41,45,47,50,70,85]- [88,99, 104,126,127,132,144,146,159,169,176,177,179,180,183,184,187,199,200,202,204]-[206, 208];

4. dynamic operations planning, which is concerned with the dynamic, minute-to-minute operations of FMS [5,8,10,17,19,30,38,61,65,72,73,78]- [80,92,93,100,105, 113,114,116,119,121,132,134]- [138,141,149,152,154,161,163,164,167,176,178,179, 180]- [183,190,202,210]- [212,214,215,217,218]

In order to identify the suitable modelling techniques for solving the problems of FMS (see Buzacott and Shanthikumar [27], Buzacott and Yao [30], Kusiak, [107], and Kusiak [108]), the literature is classified based on the following:

1. mathematical programming [5,7,8,11,13,17,20,33,36,37,50,55,73,78,85,86,92,93,99, 100,101,104,106,116,117,119,126,127,136,137,146,147,152,154,155,161,162,169,177, 181,183,184,186,200,209,212],

2. simulation [2,10,13,49,55,58,61,64,120,132,133,139,149,159,166,174,191,197,210,215]

3. queueing networks [6,34,64,65,77,191,201,210,218]

4. heuristic methods [1,11,15,39]-[42,45,47,51,53,57,70,144,161,162,182,184,199,209]

5. clustering methods [39]-[41][45,70,88,106,159,202,204,206,207]

6. control theory approach [61,72,79,80,93,164,195]

7. Artificial Intelligence and Expert Systems [44,66,76,109,111] - [114,121,135,138, 141,145,185,213].

For the purpose of reviewing the literature, an overall classification scheme is proposed (see Stecke [179] and Kusiak [108] based on (i) design of FMS; (ii) implementation of FMS; (iii) operational aspects of FMS, (iv) measure of flexibility and performance of FMS and (v) Artificial Intelligence (AI) and Expert Systems (ES) in FMS. This classification scheme is further sub-classified to bring out more pertinent factors of FMS. The details follow hereunder:

6

1. Based on the design of FMS. Browne et al. [24], Kusiak [108] and Montazeri and Gelders [132] presented the framework for FMS design. The design of FMS involves the following issues:

    (a) selection of equipment [48,53,57,64,84,106,110,117,147,155,192,209].
    (b) layout [2,59,62,75,94] -[97,118,171,191,196]
    (c) material handling [10,21,33] -[35,55,56,120,125,139,148,150,195,197].
    (d) labour issues [52,60,67,68,83,165,166,188].

2. Based on the implementation of FMS [4,12,26,46,63,81,83,128,129,131,140,142, 168,173,174,186,203]. Research reports that deal with the aspects such as precipitating circumstances, enabling conditions, technical and economic evaluations, technical and economic risks, and behavioural response in FMS come under this criteria.

3. Based on the operational aspects of FMS. Due to high capital involvement in FMS, a high rate of their utilization is important to ensure a sufficient level of return on the capital invested. Sufficient rate of FMS utilization can be guaranteed by appropriate planning, scheduling, control and monitoring strategies. The following are some of the major problems involved in the operations of FMS:

    (a) aggregate planning [20,47,86,126,127,132,176,177,180,183,187]. Aggregate planning is the intermediate-range planning, which comprises decisions related to master production scheduling and deals with a planning horizon from several days to several months in duration.
    (b) part type selection [1,2,36,85,87,99,103]. The part type selection problem consists of splitting up the production requirements for a set of part types into a number of subsets (batches) of part types for simultaneous manufacturing.
    (c) resource grouping [7,11,15,25,37,39] - [41,45,50,70,86,88,99,103,104,144,146, 159,169,179,184,199,200,202,204] - [207]. The resource grouping problem consists of partitioning the machine tools of each machine type into machine groups such that each machine in a particular group is able to perform the same set of operations.
    (d) production ratio determination [60,61,72,86,93,132,149,177,214,215]. The production ratio problem consists of determining the relative ratios at which the part types selected in problem (b) will be produced.
    (e) resource allocation and loading [5,8,13,17,19,38,51,61,65,73,78] - [81,92,100, 101,105,116,119,121,132] - [137,149,151,152,154,156,157,160,161,163,164,172, 176,178] - [182,190,202,210] - [212,214,215,217,218]. The resource allocation consists of allocating the limited numbers of pallets and fixtures of each type among the selected part types. Loading consists of allocating the operations

and associated cutting tools of the selected set of part types among the machine group subject to the technological and capacity constraints of the FMS.

4. Based on the measure of flexibility and performance in FMS [6,9,14,16,18,23,24, 27] - [29,32,46,71,89] - [91,102,108,122] - [124,130,140,143,170,173,189,193,194, 216].

    Flexibility is defined as the ability of a manufacturing system to cope with changing environments.

5. Based on Artificial Intelligence and Expert Systems in FMS [44,76,111] - [114, 138,141,145,167,213]. Expert Systems and Artificial Intelligence developed and reported in the literature occupy a crucial place in solving some of the problems of FMS.

4 Review of previous research on FMS

In the past, a number of researchers has reviewed the literature on FMS. Kusiak [108] reviewed the applications of Operational Research models and techniques in FMS and presented various design and operational problems in FMS. Buzacott and Yao [30] reviewed a number of models of FMS in order to make further improvements for the same purpose. Van Looveren and Gelders [198] presented a review of FMS planning models. Kalkunte et al. [91] presented various modeling approaches for design, justification and operational problems of FMS. Recent trends in flexible automated manufacturing were presented by Eversheim and Hermann [59]. However, most of these reviews do not account for the whole set of problems involved in FMS. Some of the reviews may even be restricted to a specific set of problems such as layout and material handling problems either with various modeling aspects or with different strategic aspects for solving problems like flexibility and its measurement in FMS.

4.1 Models based on the selection of equipments

In the equipment selection problems of FMS, generally one needs to determine the optimal number of robots, each type of machines, material handling equipments and robot assignments in every cell for balancing the production line such that the total cost of production is minimized. The importance of such research was pointed out by Kusiak [110]. He provided a framework using two integer programming models for the selection of equipments. This framework is based on the essential integration of machine tools and material handling equipments. The algorithms for part and machine selection in FMS were dealt with by Whitney and Suri [209]. Dirne [48] presented a model for estimating the throughput time in the case of manned and unmanned shift in Flexible Machining Cells (FMC) by fairly distinguishing various parameters that influence the throughput time of manned and unmanned shift. However, the author did not discuss the effectiveness of determining the run batch sizes considering the above differences. Schweitzer and Seidmann [155] presented the processing rate optimization in FMS using queuing models with distinct multiple job visits to work centers. But the number of potential models available for equipment selections and capacity planning problems of FMS is rather very limited. However, these issues are very important especially in FMS whereas the hardware are very expensive.

Lee et al. [117] constructed an algorithm for the minimum cost configuration (number of machines and pallets) problem in FMS. The work of Lee et al. [117] differs from the earlier work on solving the workload allocation in product-form closed queuing networks in that, for each possible configuration, they solved the subproblem of allocating the total workload among the machines in order to maximize system throughput. However, they did not deal with the issue of how to optimize the number of workstations and their capacity. In addition, they considered only a simple FMS where there is only one type of machine and each pallet carries only one part. Ghosh [64] presented some basic advantages of Group Technology (GT) in selecting their parts, machines and equipments. But, the author did not provide any analytical framework for evaluating the benefits of GT.

Raja Gunasingh and Lashkari [147] proposed two 0-1 integer programming formulations for grouping the machines in cellular manufacturing systems based on the tooling requirements of the parts, toolings available on the machines and processing times. An analysis of capacity planning in FMS was proposed by Tatikonda and Crosheck [192] from a case study for higher productivity and utilization. The results obtained by them seem to be interesting, but the results of a case study may not be generalized. Elsayed and Kao [57] presented a method for determining the optimal number of machines for each operation, and the optimal number of robots required in each cell by minimizing the total production cost per unit. However, they did not include the schedule of the material handling equipments in their modeling which in fact influences the speed of the machines or cells.

4.2 Models based on layout design for FMS

The problems relating to layout design in FMS is one of the foremost important issues and this should be resolved suitably at the beginning of the system design. Since the hardware used in FMS is rather expensive, the FMS layout designer should select suitable layouts considering the various alternative layouts. The machine layout problem implies the arrangement of machines on a factory floor so that total time required to transfer material between each pair of machines is minimized. Apart from time and distance factors, factors such as handling carrier path, clearance between machines, etc. are to be taken into account while evaluating the alternative layouts. Kouvelis and Kiran [96] reviewed the recent results on FMS layouts for identifying the important aspects to be considered while designing the layout for FMS as compared to that of conventional manufacturing systems. However, the FMS layout literature focussing on analytically modeling the problem and developing solution algorithms is very limited.

Recently, a survey of solution procedures for facilities layout problems has been reported by Levary and Kalchik [118]. Nevertheless, while applying reported methods or tools in the literature, one has to account for the characteristics of FMS such as generally unequal machine sizes, interaction between layout decisions and queuing performance measures of an FMS, and flow control issues, including the interaction between processing requirements, travel times, part mix and process selection. Heragu and Kusiak [75] discussed the issues concerning the application of the Quadratic Assignment Problem (QAP) formulation in a certain FMS implementation. However, QAP formulations do not account for the interaction between layout decisions and queuing performance measures of an FMS. Solberg and Nof [171] demonstrated the significance of such interactions with a view to arrive at a suitable layout for FMS. The effects of layout decisions on throughput rates (via transportation times) were captured in a formulation of the FMS layout problem by Kouvelis and Kiran [95]. Furthermore, Kouvelis and Kiran [96] developed single and multiple period layout models that incorporate the queuing and product mix uncertainty aspects of FMS layout decisions.

Storage location for WIP in FMC with handling robots under discrete and continuous space assumptions was studied by Tansel and Kiran [191]. Gaskins and Tanchoco [63] proposed an integer programming formulation for determining the optimal flow path that minimizes total travel distance of loaded Automated Guided Vehicles (AGVs). Kiran and Tansel [94] considered the optimal location of a pick up point on a material handling network and developed a polynomial time algorithm for determining the optimal pickup point location. Kouvelis and Lee [97] presented two different formulations of the MHS layout problem. Later on, attention has been focused on analyzing the specific layout types that are to be implemented in FMS. The main difficulties with these methods are that they consider only one criterion at a time and ignore many other operational issues [95,118,171].

Moreover, a number of researchers has attempted to develop methodologies which allow the human decision maker to play an interactive role in the facility layout analysis problem. The purpose has been to integrate quantitative and qualitative objectives and enable the decision maker to manipulate resultant layouts to include factors that are not captured by the model. It has been indicated that the unidirectional and unicyclic material handling system layout are preferred to other configurations because of their relatively lower initial investment cost and higher material handling flexibility. But at the same time, it should also be observed that the flexibility of the material flow is rather limited and this may have some counterproductive effect on production rate and hence on productivity. Therefore, suitable trade-offs may be helpful while evaluating the different alternative layouts. Afentakis et al. [2] presented a method to solve the physical layout problems for higher utilization of FMS and studied the impact of material handling systems on layout design. However, they assumed that the process planner selects only one routing for each part type which is against the concept of flexibility.

4.3 Models based on the Material Handling System (MHS)

In the traditional MHS, a human element is involved in the transportation of materials between various locations, but in FMS there is a less human intervention. This aspect has been fully supported by the developments of Automated Guided-Vehicle Systems (AGVS) and computer-controlled MHS. A dynamic material handling for a class of FMS was presented by Cassandras [35]. Earlier, Cassandras [34] offered an autonomous material handling in computer integrated manufacturing. Egbelu and Tanchoco [55,56] investigated the automatic dispatching rules and potentials for bidirectional guide path AGVS. Moreover, Raman et al. [148] discussed the simultaneous scheduling of material handling devices in automated manufacturing. The effects of the number of jobs allowed into an FMS on its system performance and the relative performance of different machine and AGV scheduling rules were studied by Sabuncuoglu and Hommertzheim [150] with help of a simulation model.

Ashayeri and Gelders [10] offered an interactive Pascal program which generates the GPSS-PC models for designing efficient and economical automated material handling systems. A tandem configuration for reducing software and control complexity of the AGV-based MHS was proposed by Bozer and Srinivasan [21]. Ozden [139] performed a simulation study on a small FMS to consider simultaneously the design parameters such as traffic pattern, number of AGV and the carrying capacity of each. AGV, and queue capacity of the machines. More precisely, the use of a computer simulation reduces the risk in the design of MHS and improves throughput in automated material handling operations (Trunk [197]). Mahadevan and Narendran [120] addressed the issues involved in the design and operation of AGV-based material handling systems for an FMS. Thomasma and Hilbrecht [195] presented specification methods for material handling control algorithms in FMS.

4.4 Based on labour issues in FMS

Labour issues in FMS (capital intensive) differ from those for conventional manufacturing systems (labour intensive) in the following aspects: (i) there are no machine operators for individual machines; (ii) the set-up operations are performed by the machine itself or by robots; (iii) there is no manual material handling system; (iv) layout job is supervisory in nature; and (v) the labour is multi-skilled in one or more tasks (Graham and Rosenthal [67]; Dunkler et al. [52]). Despite the increasingly important role of the human-computer interface in FMS, it is not clear which functions should be allocated to the human (supervisor) and which should be allocated to the computer. However, the issue of how the arrangement between human and computer in a particular decision control activity should be designed is yet to be resolved. Sarkar [153] presented a model for estimating the numbers of workers required in an assembly process, but the model has to be modified in order to apply it in an FMS environment considering the labour characteristics of FMS.

Fazakerly [60] examined the human aspects of GT and cellular manufacturing. Since labour unavailability and machine breakdown in FMS leads to a heavy loss in production, multi-skilled labour force is preferred for FMS. The human aspects of FMS have been discussed by Gupta [68] and Sharit [165]. However, they do not optimize the number of supervisory personnel required to take care of the machines and the number of machines, an operator can look after the loading, unloading, routine maintenance and other minor settings. Sharit and Elhence [166] discussed the limitations of both human and computer in achieving both of these system performance objectives. But they did not offer any concrete framework for overcoming these limitations. Suitable methods for modeling the interaction between human and computer system in FMS will facilitate the efficiency of the coordinating activities.

4.5   Models based on the implementation of FMS

The literature that deals with various issues of implementing the FMS including economic and technical justification for FMS is presented in this section. A conceptual model for the implementation of new manufacturing technologies has been developed by Avlonitis and Parkinson [12] . Gelders and Ashayeri [63] studied various management issues in flexible process manufacturing. One of the most important problems in the implementation of FMS is to measure the performance of different alternatives.

Several studies have been conducted on the economic justification over the last few years, especially after the development of FMS (Wallace and Thuesen [203]). The limitations of many of these traditional approaches, in the context of flexible automation technologies, have been bestowed in recent years. Monahan and Smunt [131] presented a stochastic dynamic programming approach, considering five factors: interest rates, technological levels, product mix, flexibility and inventory costs. Nonetheless, an inadequate treatment of the challenge from current and FMS technologies with market characteristics, and so on, are not considered in the traditional models. Suresh and Sarkis [186] developed a mixed integer programming formulation for the phased implementation of FMS modules.

Son and Park [174] measured the flexibility monetarily using computer simulations. They considered four nonconventional costs for measuring the flexibility such as set-up, part waiting, equipment idle and inventory. The opportunity costs can give manufacturers a valuable information not only for evaluating these systems performance (Son and Park [173]), but also for decision making about other advanced manufacturing systems (Park and Son [140]). However, assumptions such as: only three parts need to be processed, seven different tools are used, only one tool is used for a part in a process, and poison demand of parts, may limit the application of the model to generalized FMS. Burstein and Talbi [26] proposed a mixed integer programming approach with linear integer variables and non-linear continuous variables that addresses the problem of the challenge from upcoming best-practice technologies to the existing system.

In recent years, a number of research articles [140,142,185] has focused its research on the cost justification of manufacturing investments, particularly on the issues of investing in new manufacturing technology such as FMS. The dramatic changes in manufacturing technologies force to re-evaluate traditional cost accounting and investment analysis systems. Pollard and Tapscott [142] discussed an investment methodology for evaluating new manufacturing technologies. This method offers a systematic and rigorous process for analyzing a long-term portfolio of manufacturing investments such as new information systems, factory automation, FMS, and CIM. Suresh [185] presented a Decision Support System (DSS) structure for flexible automation investments. More significantly, appropriate models at different stages of economic evaluation of FMS need to be developed with the objective of increasing the use of the proposed DSS.

4.6 Aggregate Production Planning (APP) models

Generally, aggregate planning models of FMS deal with estimation of number of machines, number of pallets and fixtures required, and the grouping problem. At the aggregate planning level, Stecke [183] argued that a single closed queuing network model, considering multiple server queues (machines), can be used to solve the problems of FMS. However, assumptions concerning the arrival and departure rates are questionable with respect to a specific situation. Stecke and Solberg [176] presented an optimal planning of computerized systems - the CMS loading problem. Stecke [180,183] discussed in detail the problems of APP in FMS. However, most of the models do not include the cost aspects of the resources and optimization of the facilities required.

Recently, Mazzola et al. [127] presented a hierarchical production planning model which integrates FMS production planning into a closed-loop MRP situation. Their FMS/MRP production planning framework consists of planning, grouping and loading, and detailed scheduling problems. Later Mazzola [126] explored the necessary heuristics for the FMS/MRP rough-cut capacity planning (FMRCP) problem by systematic splitting of planning batches. The advantages of using this method are to take into account the priorities set by the MPS and MRP systems and the limit on the available capacity of FMS. However, the assignment of parts in a subset to the same set of facilities lead to a fixed route of part flow and this is against the notion of flexibility of the systems. Chung and Lee [47] suggested a heuristic algorithm for scheduling FMS. Hwang [86] and Stecke [177] presented different production planning problems and appropriate models for solving those problems in FMS. Boctor [20] presented an efficient heuristic for the machine grouping problem in Flexible Machining Cells (FMC). He also presented a new linear zero-one formulation which seems to have most of the advantages observed in other models. However, most of these models consider only a part of real-life FMS; without considering realistic system configuration and other operational features of the systems.

4.7 Part-selection problems in FMS

In practice, however, if the number of part types is extensive and/or the number of operations required by each part is large, achieving an improvement in performance is difficult using the FMS. This is due to the constraints included in the design of the system such as the number of tools of a particular type on hand or the number of pallets available is not sufficient to process all parts simultaneously. This problem can be solved by partitioning the part types into groups (families) so that the parts in each group will be processed simultaneously. Thus, it leads to the part type selection problem in FMS.

In the past, a number of part selection methods has been reported in the literature emphasizing on selecting similar parts. Several similarity measures and their associated clustering or heuristic methods were presented (Kusiak [103]; Chakravarty and Shtub [36]). However, these methods do not consider many realistic constraints such as tool magazine capacity, pallet capacity, availability of tools, etc. present in the part selection problem. In addition, the similarity measures are often subjective, consisting of a mix of nominal and numerical data. Therefore, further research is needed to include such constraints in part-selection methods. However, in order to arrive at the optimal selection of parts, these methods should include the equivalent quantitative factors which perhaps based on subjective factors. Kumar et al. [99] discussed the problem of grouping of parts and components in FMS. The actual grouping is done by modelling the problem as a k-decomposition of weighted networks. Hwang and Shogan [85] discussed a part selection problem for an FMS with general purpose machine tools but with no tool transportation devices. They proposed a maximal network flow model with two side constraints. However, this model ignores the tool overlapping as considered by Stecke [179], whereas the selection of part types depends upon tool availability.

Hwang and Shanthikumar [87] proposed a model for production planning in FMS. However, this model does not account for the constraints on tool availability in the system for processing a group of parts in a particular planning period. Afentakis et al. [1] focused on batching approaches to the problem of part type selection in certain types of FMS. They offered multifit and contraction heuristics for solving part selection problems and tested them under a number of problem conditions. However, the application of these heuristics is limited to only cyclic scheduling policy and to the situation whereas only one unit of each part type is needed to produce one unit of the assembly. Afentakis et al. [2] described a part-type selection problem considering dynamic layout strategies in FMS. But the situations regarding FMS layout do not permit so easily to change the layout and hence this study needs to be evaluated considering the implications of various relevant costs.

4.8 Models based on resource grouping

The main objective of GT in FMS is to reduce set-up times (by using part-family tooling and sequencing) and flow times (by reducing the set-up and move time). A major step in designing an FMC is grouping of parts into families and the corresponding machines into manufacturing cells. GT provides a coding and classification scheme for grouping various parts and products with similar design and/or machining processes into a family of parts and corresponding machines into machine cells. Rajagopalan and Batra [146] and Purcheck [144] introduced a significant shift in the direction of thinking about GT techniques. Recently, Chandrasekharan and Rajagopalan [40,41] and Chan and Miller [39] presented a number of approaches based on GT for grouping of parts and machines, which perhaps enriched the literature on GT application. However, these methods do not include the demand for parts and other processing parameters such as part type, sequencing rules, operation time per unit, etc. in machine-component grouping.

Steudel and Ballakur [184] presented a heuristic based on a dynamic programming for machine grouping in a manufacturing cell formation. Waghodekar and Sahu [202] presented a machine-component cell formation using GT. The approaches and models available for solving the grouping problem have been classified into three major categories: GT (Chakravarty and Shtub [38]; Hyer and Wemmerlov [88]; Kusiak [104], sequential decision procedures, and mathematical programming (Hwang [86]; Kumar et al [99]; Kusiak [103]; Stecke [179]. But very little research has been aimed at comparing the numerous cell formation procedures. In general, GT determines part clusters by measuring 'similarities' among parts based on a coding scheme. For the FMS grouping problem, the GT approach needs to code the characteristics of the parts and their associated components such as tools, pallets and fixtures. Apart from this, GT has to consider the limits on the number of part types and associated components in each group.

The problem of group formation and the solution procedures have been reviewed by Vannelli and Kumar [199] and Ballakur and Steudel [15]. Kusiak [103] used Lagrangian relaxation to solve a 0-1 linear integer programming formulation of the grouping problem which considers only similarities of the objects (parts) for clustering purposes. Stecke [179] formulated the machine grouping as a nonlinear mixed integer programming problem. Instead of using the concept of similarity as a criterion to group parts into families, Hwang [85] developed a model to maximize the number of part types in a batch for simultaneous processing considering the capacity constraints of tool magazines in FMS. However, he did not include other constraints such as due dates and part quantities in his formulation. Kumar et al. [99] modelled the machine-part grouping problem as an optimal k-decomposition of weighted networks, which is a 0-1 quadratic programs. Shtub [169] demonstrated that the simple cell-formation problem is equivalent to the GAP. The major draw back of this model is that it does not consider the geometric shape in the similarity measures except that of a process plan. Gupta and Seifoddini [70] presented a similarity coefficient based framework to take into account product demand and other processing parameters in machine-component grouping. This similarity coefficient assumes that product demand remains the same over a particular planning period. However, a change in demand pattern may require different grouping of machines in each cell.

Al-Qattan [7] developed a new method of forming FMC, based on branching from a seed machine and bounding on a completed part. This method treats the formation of machine cells and part families as a network analysis problem. Vanneli and Kumar [199] have obtained similar FM cells for the same problem, using graph theory, but the method proposed by Al-Qattan [7] gave more alternative solutions. The literature available on GT is not too realistic considering that over a time period. For instance, the variation in product mix, product design, market situation and other technical factors lead to a reorganization of the whole system. Ventura et al. [200] formulated the problem of grouping parts and tools as a 0-1 Linear Integer Program (LIP) and this formulation is equivalent to the quadratic integer program of Kumar et al. [99]. Chu and Tsai [45] compared three array-based clustering, viz. algorithms-rank order clustering, direct clustering analysis, and bond energy analysis (BEA) for manufacturing cell formation. According to their experiments, BEA performs better than the other two methods, without considering the type of measures or data set used. Divakar et al. [50] considered a generalized group technology problem of manufacturing a group of parts in which each part can have alternative process plans and each operation in these plans can be performed on alternative machines. However, considering the integer programming capability in modelling the problem, there is a need to develop heuristics to solve the problem. Askin and Chiu [11] developed a mathematical model and heuristic for the GT configuration problem incorporating for the first time the costs of inventory, machine depreciation, machine set-up and material handling into a mathematical programming formulation. However, comparatively few authors have considered the cost aspects in their cell formation procedures.

4.9 Models based on the production ratio determination

A number of analytical and simulation models have been reported in the literature [86,177] which deal with the problem of determining the relative production ratios of selected parts. Han and McGinnis [72] presented a flow control method for an FMC to minimize the stockout costs in meeting time-varying demands from downstream cells. The same problem was studied by Kimemia and Gershwin [93], using an optimal control concept in case of constant demand rates and random station failures. However, if the length of a control period is fairly large, control will be inordinately reactive to changes in the environment, thereby causing an operational instability. Besides, transportation time has been neglected in determining the production ratio in FMS.

On-line control and scheduling have generally been used in FMS. For instance, Wu and Wysk [215] described a control mechanism of job dispatching heuristics using the results obtained from a simulation. Also, the authors [214] offered a multi-pass expert control system for controlling the FMC. Most of the articles on FMS with random processing times have employed queueing models (Buzacott and Yao [30]). However, the concept of OPT can be employed while estimating the production ratio in FMS. It is well expected that the production ratio should match with the speed of the automated MHS considering the loading/unloading time of the parts. In this respect, the determination of production ratio should be treated as an integral part of decision making with AGVs. Ro and Kim [149] presented a multi-critiera operational control rules in FMS using simulated results. The special feature of their approach is to consider the scheduling of AGV. The process of batch splitting and forming will result in an efficient flow of materials in FMS. Also, constraints such as operator availability, machine breakdown, etc. have to be incorporated in the determination of production ratio in FMS.

4.10 Models based on resource allocation and loading

Resource allocation includes the allocation of a limited number of pallets and fixtures of each type among the selected part types. The loading problem is concerned with the allocation of part operations and required tools amongst machine groups for a given product mix. It is important to observe that loading decisions in FMS are constrained by the number of factors such as the number of tool slots available on the tool magazine of a machine spindle, the number of slots a tool occupies on the magazine, non-splittting of jobs, capacity of various machines, and so on. Owing to high cost of capital investment in the FMS, the performance measures are quite complex, and often involve multiple objectives. In addition, these objectives may differ from system to system depending upon system configuration, types of parts to be produced, demand conditions, etc. (Stecke [179].

Sharifnia et al. [164] developed a method for flow control of parts in a manufacturing system with machines that require set-ups. The basic idea employed in their method is to use the production rate targets determined at the higher level for generating the set-up schedules at the lower level. Hildebrand [77] presented the scheduling of machines in Flexible Machine Systems wherein the machines are subject to failures. Chakravarty and Shtub [38] presented an analysis for capacity, cost and scheduling in a multi-product FMC. Kimemia [92], Maley et al. [121], Nof et al. [135], Sarin and Chen [152], and Sawik [154], and Stecke [176,181,182] presented a number of models and solution procedures for operational control problems of FMS. Most of these authors assumed that there is no batch splitting, and breakdown of MHS and equipment in the FMS. Hintz and Zimmermann [78] offered a fuzzy linear programming to control the releasing of parts into FMS and the scheduling of parts and tools. Solot [172] dealt with a new concept of controlling the operations of an FMS. This concept is based on the integration of planning and scheduling problem using the joint modelling of Operations Research and Experts Systems.

In a random FMS, the loading decisions are usually dynamic when the product mix is not clearly defined at the time of installing, the system is arduous. For a dedicated FMS, Stecke [179] formulated the loading problem as a 0-1 nonlinear mixed integer program. The relevance of FMS loading models to the generalized transportation model and generalized assignment model has been brought out by Kusiak [105]. Stecke [183] presented a hierarchical approach for solving the problem of machine grouping and loading in FMS. Greene and Sadowski [65] offered a mixed integer programming formulation for loading and scheduling of multiple FMC. However, they assumed that each cell is independent but receives its jobs from a common buffer, and viewed the loading problem as selecting a subset of jobs and allocating jobs among machines. Yao and Buzacott [217,218] concerned themselves with the state dependent routing problem and models with limited local buffers for FMS. Shankar and Srinivasalu [162] developed a two-stage branch-and-bound procedure with the objective of maximizing the assigned work load in an FMS. They also presented heuristic procedures with the bi-criterion objective of minimizing the work load imbalance and maximizing the throughput for critical resources such as the number of tools slots on machines and the number of working hours in a scheduling period.

Hitomi et al. [80] solved the design and scheduling problems for FMS as a two-machine flow-shop problem with finite buffer space and automatic set-up equipment (Hitomi et al. [79]). They proposed that the set-up and machining operations can be simultaneously conducted on an index-pallet changer which has a multiple number of clamping devices, which serve as centering, machining and buffer stations (e.g. Seidmann and Schweitzer [157] ; Seidmann and Nof [156]). Stecke and Solberg [178] and Stecke [179] provided the foundation for the approach for solving the problem of resource allocation and loading in FMS. However, they did not consider the aspects of refixturing and limited tool availability in the loading problem of FMS. Realizing the practical importance of this, Lashkari et al. [116] suggested a non-linear program in (0-1) integer variables for the loading problem. Wilson [212] reformulated the problem of operation-allocation to include the aspects of refixturing and limited tool availability. The literature concerned with the design and operational problems of robots in FMS is comparatively small, although it plays a vital role in the system. Acknowledging the importance of these problems in FMS, Blazewicz et al. [19] presented algorithms for scheduling the robot moves and parts in a robotic cell. Sarin[151] treated the computerized analysis of a robotized production cell.

Operational changes using different scheduling rules which are based on the characteristics of the system may further reduce overall machine set-up time in cellular manufacturing systems. At the same time, these rules will eliminate some of the deficiencies while adopting a variety of parts. We and Wysk [215] described a scheduling algorithm which employs discrete simulation in combination with a simple part dispatching rule in a dynamic fashion. Various scheduling approaches and algorithms are presented by Shalev-Oren et al. [160], Shanthikumar and Stecke [163], Stecke [180], Vinod and Altiok [202], Kumar [100], O'Grady and Menon [136], Wilhelm and Shin [211], and Wu and Wysk [214]. However, little research has been reported in group scheduling of FMS. In addition, the literature available on FMS dealing with disturbances such as machine breakdown is rather limited. Considering the importance of such disturbances, especially in FMS, a queuing model was presented by Widmer and Solot [210] to include machine breakdown and maintenance. Ro and Kim [149] presented three new process selection rules (alternative routings directed dynamically, alternative routings planned directed dynamically, alternative routings planned and alternative routings planned and directed dynamically) in FMS with limited local buffers. Montazeri and Van Wassenhove [133] analyzed the characteristics of a general-urpose user-oriented discrete-event simulator for FMS. However, one has to examine whether the simulator is compatible with the type of system to be investigated in terms of system configuration and operational aspects.

Garetti et al. [61] studied the impact of product mix and characteristics of the system on the performance of loading and dispatching rules with the help of a simulated FMS. The most common approach which subdivides the loading process into two phases (Nof et al. [134] and Shanker and Tzen [161]): The first phase is loading, consistent in the sequencing of the jobs at the entrance of the system so as to make them accessible for production. The second phase is dispatching, consisting in the choice of jobs to be loaded from those waiting to be produced by a machine. Moreover, there are further interesting solutions (Doulgeri et al. [51]; O'Grady and Menon [137]; Stecke and Solberg [178]) for various configurations of FMS. Kumar et al. [101] combined the problem of grouping and loading in an FMS by formulating it as a multi-stage multi-objective optimization model. Han et al. [73] analyzed the effect of tool loading methods, tool return policies, queue formation methods, and job scheduling rules using simulation.

Akella et al. [5] developed a linear programming model for computing a quadratic approximation to the value function. However, the assumption that each part has a single route through the system is in fact in contradiction with the existence of alternative routings in FMS. Berrada and Stecke [17] treated the loading problem of FMS with the objective of balancing the workload using a part-movement policy. A work-center loading problem in flexible assembly with double objectives, such as workload balancing and part movements was studied by Ammons et al. [8]. Tang [190] proposed a job scheduling model which would minimize the number of tool changes at a single machining center. Liang and Dutta [119] developed a mixed integer programming model for the machine loading and process planning problem for a process layout environment. This model relaxes the most commonly used assumption that each operation can be assigned to only one machine. However, they did not include the aspects of breakdowns, random job arrivals, facility layout and reallocation, etc. Yih and Thesen [219] presented a class of real-time scheduling problems and showed that these can be formulated as semi-Markov decision problems. But the limitation of the semi-Markov process is yet to be explored in detail in order to model a real-life FMS.

4.11 Based on the measure of flexibility and performance in FMS

System flexibility is the core subject of FMS concepts and practice. But this concept is not well defined. This is attributed partly to its incoherence: different researchers have emphasized different types of flexibility only on a specific aspect of the total system. However, there have been some research reports to analyze the system flexibility of an FMS from a descriptive point of view rather than an analytical point of view. Chung and Chen [46], for example, presented total system flexibility, along with several routing flexibility measures, to assess the value of system flexibility in the FMS environment. The assessment is based on the capability of an FMS in cushioning the effect of change in manufacturing environments. Browne et al. [24] and Buzacott [28] attempted to divide the general term 'flexibility' into a number of elementary concepts.

Despite this, there has also been a number of attempts to define the term 'flexibility'. Buzacott [27], Browne et al. [24], Swamidass [189], and Kusiak [108] proposed various types of flexibilities. In summary, all these types fall into two categories: machine level flexibility and system management level flexibility. Alberti and Diega [6] presented a new measure for evaluating the operational performances of an FMS. This was in order to integrate the economic issues in the management of production environments defined by high-level automation, flexibility and integration. The performance evaluation of FMS with blocking was discussed by Tempelmeier et al. [194]. A number of problems, however, are still unsolved. The main problem (Blackburn and Millen [18]; Kalkunte et al. [91]) is the difficulty of measuring economic benefits which result from flexibility. Azzone and Bertele [14] outlined a method for the evaluation of FMS which considers both economic and strategic aspects.

According to Chung and Chen [46], it is evident that product flexibility described by Browne et al. [24] is called variant flexibility and high-volume/low-variety flexibility in Swanudass [189]. On the other hand, job flexibility described by Buzacott [28], which is called process flexibility in Browne et al. [24], differs substantially from the job flexibility defined by Kusiak [108].

Moreover, since the key term in FMS is system, in order to better evaluate the value of flexibility for an FMS, there is a demand to develop a system flexibility model from a holistic viewpoint. Recently, Gupta and Goyal [69] classified the literature available on the concepts and measurements of FMS. Other researchers who deal with flexibility include Barad and Sipper [16], Carlsson [32]. Mills [130], and Slack [170]. Park and Son [140] developed an economic evaluation model for advanced manufacturing systems.

Many authors ([23,24,189], etc.) during the last few years attempted to attain a qualitative understanding of flexibility in manufacturing systems. A number of them have aimed to attain a quantitative understanding of flexibility in manufacturing systems (Buzacott [28] and Browne et al. [24]). Brill and Mandelbaum [23] and Mandelbaum and Brill [124] presented a framework for defining measures of flexibility in production systems. The framework has been given in general terms, so that a user will be able to analyze specific situations within the general framework. Some authors have emphasized the importrance of management, information and learning in the use of flexibility in manufacturing systems (Jaikumar [89,90]). The relationship between flexibility and productivity has been studied extensively in the literature (Gustavsson [71]; Buzacott and Mandelbaum [29]; Mandelbaum and Buzacott [122,123]). Decision-theoretic approaches have been used to study a wide range of issues in flexibility and general decision making (Mandelbaum and Buzacott [122,123]. Arbel and Seidmann [9] and Kumar [102] discussed the various measures of flexibility and different methods to measure them. An entropy-based measure for flexibility has been proposed by Kumar [102]. Examples of production systems which illustrate a measure-theoretic approach for measuring flexibility are given in Brill and Mandelbaum [22]. Taymaz [193] offered different approaches to the flexibility. However, the application of these approaches are limited as they were experimented with only one machine.

4.12   Artificial Intelligence and Expert Systems in FMS

Generally, an FMS is expected to operate as an intelligent autonomous system that delivers products according to a dynamic strategy. However, achieving autonomy and intelligence in an FMS requires suitable co-ordination between scheduling and control tasks. The need for designing real-time control systems with a high level of intelligence has been emphasized in FMS (Kusiak [109]). The significance of utilizing

AI in providing the level of intelligence required by the scheduling function and the facilitation of the integration of scheduling with control has also been pointed out.

Recently, a number of software systems incorporating the features of AI has been established. The high-quality performance of these systems, which is too complex for conventional programming techniques, has given rise to the term Expert Systems. Kusiak and Chen [111] reviewed the research and application of Expert Systems in production planning and scheduling. However, very few ES have been developed for solving the problems of design, justification, and operation in FMS. In this direction, Rabelo and Alptekin [145] presented the design and implementation of an intelligent scheduling system in an FMS scheduling/control architecture using AI technologies. Chryssolouris et al. [44] discussed a decision-making framework for manufacturing systems which pertains to the problem of allocating the system's resources to manufacturing tasks, Park et al. [141] proposed a framework incorporating machine learning into the real-time scheduling of an FMS.

In the past, the hardware part of FMS received great attention and improvement in terms of flexibility. However, software availability has not reached the same level of flexibility as that of the hardware. In this direction, O'Grady and Lee [138] designed an intelligent cell control system for automated manufacturing. Shaw [167] applied inductive learning to improve knowledge-based Expert Systems and a pattern-directed approach to FMS scheduling. Knowledge-Based Machine Layout (KBML) Kusiak and Heragu [112]) is an ES for machine layout in automated manufacturing systems. However, it is assumed that the manufacturing system is automated and predefined, that is, the degree of flexibility has not been defined in their knowledge base. Nonetheless, the knowledge-based approach has received very little attention from researchers as compared to that of optimization in scheduling problems. Realizing the significance of such systems, particularly in FMS, Kusiak [103,104] designed a knowledge-based scheduling system for an automated manufacturing environment. Wu et al. [213] designed a cellular manufacturing system using a syntactic pattern recognition approach. Heragu and Kusiak [76] presented KBML designed to solve the machine layout problem. KBML combines the optimization and Expert System approaches and considers quantitative as well as qualitative factors while solving the machine layout problem. Moreover, KBML allows changes of the solutions which are not implementable.

5 Limitations of the FMS literature and future research directions

Integration between theoretical and practical issues is predominant in today's FMS environment. Nevertheless, a clear distinction between the two aspects is usually observed both in practice and research studies; a kind of gap exists between the two fields. The application of the existing FMS models is generally limited due to the following: (i) FMS configuration considered in modeling; (ii) assumptions made in the model; (iii) modeling technique and its capability; and (iv) solution procedures employed and their efficiency. The following sub-sections present some of the limitations of the available literature on FMS and future research directions.

5.1   Selection of equipments

The research reported on equipment selection problems is scant though it has a very important role in FMS. The basic objective in the equipment selection is that the parts are to be produced quickly, with the secondary objective being higher utilization of the facilities. In addition, there are more specific aspects that are to be considered in the selection of equipments: (i) nature of the material flow and bottle-neck operations; (ii) production lot-size and scheduling, (iii) feasible production cycle time per pallet; (iv) achieving JIT production in FMS; and (v) staffing level and the shifts required to meet the production goals.

Furthermore, economic investment analysis and technical feasibilities are to be given due consideration while evaluating different alternatives in equipment selection problems.

The available literature on equipment selection in FMS has mainly treated problems with respect to machine tools without considering the selection of pallets, tools, storage systems, robots and, related softwares and hardwares. However, the system design well depends upon the number of robots, pallets, fixtures and storage systems. Further, the design aspect of MHS has not been given due consideration as it really deserves significant attention in the equipment selection problems in FMS. The synchronization of MHS with respect to the processing rate of the stations is very important and the failure to do so may even result in decreased productivity. Apart from this, the constraints on the tool magazine capacity and number of pallets are also to be taken into account in the problem formulation in order to arrive at the correct optimal solution. Cariapa [31] suggested that multimode machine tool increases the effectiveness of FMS with the help of suitable criteria.

In the selection of equipments, here are some of the issues in FMS that are to be considered for further investigation in terms of modeling and application: (i) the bottleneck at the load/unload stations and at robots; (ii) use of dedicated fixtures with different numbers of parts per pallet balances the work center loads and reduces variation in flow times; and (iii) the selection of the actual production rates to be used under different operational scenarios. The application of software tools and queuing optimization models have an important role in the generation and quick evaluation of different design and operational decision options. Applications of queuing theory and a rough-cut-capacity planning approach can be used to solve the problems of equipment selection along with simulation studies and pilot projects. Further, suitable criteria are to be established in order to be successful with equipment selections in FMS.

5.2   Layout design

In FMS, layout decisions cannot be treated as independent as they interact with the following decisions: (i) the number and capacity of the stations; (ii) the number and capacity of the storage units; and (iii) the MHS design. Although there are several studies suggesting a relation between these layout decisions, there is no comprehensive study to identify the level of these relationships. Hence, this research area needs due attention from production researchers. Also, the impact of part flow on FMS layout decision needs to be studied in detail. Consideration should be given to MHS while designing the layout of FMS apart from the usual attributes such as width of the guided path, unidirectional/bidirectional MHS, robot movements, distance between various facilities, storage units, load to be transported, etc.

Since layout decisions in an FMS interact with APP, a study involving the sensitivity of the layout decision to these would answer the question how to split the design problem in an FMS. The balancedness property is extremely useful in developing more efficient models and solution techniques, especially when the balancedness is combined with other special structures, and limited part flow and storage capacity (for more detail, refer to Kouvelis and Kiran [96]). In addition, the buffer capacity and location of the storage unit are also important factors in determining layout of FMS. Hence, a different approach is required to integrate storage capacity decisions required to integrate storage capacity decisions with layout decisions. The following situations which relate to the distance metric used in layout models are to be accounted while designing the layout for FMS; (a) loop layout problem, (b) line layout problem, (c) multiple loop problem, (d) layout, storage and MHS selection, and (e) optimal system configuration.

5.3 Material Handling Systems

The following are some of the parameters that are frequently considered in the design of the material handling systems: (i) traffic pattern; (ii) number of AGV; and (iii) queue capacity of the machines. Generally, the objective considered in the design of MHS is to minimize the travelling time of the vehicles. However, the movement of the MHS also depends upon the parts loading and unloading systems and the size of the storage systems. Therefore, simple heuristic rules are needed for studying these situations.

In most of today's FMS, mainly AGV are used for material handling purposes. Therefore, efficient scheduling rules are very important for AGV in relation to the speed of the machines for improved performance of the system. For instance, use of robots necessitates presence of enough space for the movement and synchronization of loading and unloading operations of robots along with the speed of the AGV. Furthermore, synchronization of AGV and FMC can be treated as a critical source; accordingly, one can establish the level of various operational parameters. Simulation can be used for estimating the value of these parameters. The following assumptions can be avoided in modeling the problems of MHS:

1. machines and AGV have the unit capacity;

2. no pre-emption or breakdowns; and

3. tooling and pallet availability are not considered.

The problems of multi-vehicle systems and strategies for resolving them are to be examined with the help of analytical and simulation models.

5.4   Labour issues

A number of research reports have appeared in the literature to deal with various human factors such as education, training, and ergonomics in conventional manufacturing systems. In contrast, labour issues in FMS have been simply ignored except there are very few research reports in the literature. Also, while designing the robots and their movements, and other Artificial Intelligence sub-systems in FMS, parameters like human factors, especially, the interface between human and computers plays a significant role in establishing various necessary autonomous sub-systems of FMS.

Generally, a conventional production system estimates the number of workers/ operators required based on the number of machines available, supervisors and work force for material handling, workload per operator, etc. The same approach may not be applicable in the case of FMS. For instance, a skilled laborer who has been trained for FMS can look after more than one machine as most of the machines are automatic in nature. This implies that he/she only has to supervise a number of machines, and perform tasks such as loading and unloading of parts. Further, this situation may not be applicable for flexible assembly and fabrication systems, and perhaps they may need different criteria to select the number of flexible operators/workers required. Also, ergonomic aspects have been ignored in determining the workload and level of skill required considering the automation level in FMS. The number of machines an operator can look after has not been investigated in detail. Further, the interaction between human, robots and computers needs to be researched further in relation to FMS. Appropriate models need to be developed to determine the optimum number of operators, training hours, work load per operator, etc., in FMS.

5.5   Implementation issues

More basic problems in FMS, like the inadequate integration between corporate and manufacturing strategies, have contributed to slow adoption rates. Under these circumstances, the problem of incremental implementation and integration of CNC technologies needs to be discussed well in advance. The conventional costing techniques employed in traditional discrete production do not apply to an FMS environment. For example, activity-based costing techniques may be suitable for FMS. Further, suitable management accounting principles have to be established for evaluating different alternatives in the implementation of FMS. The literature available on the problem of multi-machine replacements is relatively sparse though these replacements prevail in many real-life FMS. Hence, this issue should be given due attention for further improvements. In particular, the following assumptions limit the application of FMS implementation models and approaches:

1. single machine tool replacements;

2. static part families;

3. single-valued input estimates

Therefore, it emphasizes the need for pilot studies along with simulation analysis. Further, the implementation decisions should also include the market characteristics.

5.6 Aggregate production planning

Most of the reported APP models attempt to balance the workload in order to improve the operations of the FMS. However, there are some conflicts in FMS while using only the balancing of workload as a performance criterion for efficient operation of the whole system, since the performance of an FMS depends upon a number of other criteria as reported in [183]. Therefore the multiple criteria decision making principle can be utilized for the purpose of APP. This particular problem is very important in FMS as this decides the future performance of the whole system; therefore, simulation of a pilot study would facilitate future decisions on FMS. More notably, Stecke [183] suggested an hierarchical production planning method using queuing analysis at the aggregate level and using mathematical programming and heuristics at the disaggregate level in FMS. Montazeri et al. [132] suggested that simulation is an efficient tool to verify design concepts, to select machinery, to evaluate alternative configurations and to test the system control strategies of an FMS. The following assumptions could be removed by suitable modeling and heuristics:

(a) a machine can process only one piece of a part at a time; and

(b) every tool occupied only one slot on the tool magazine of a machine center.

Simulation can also be used to avoid these assumptions where they are not applicable in FMS. Application of transfer line models to FMS may lead to some significant results concerning the planning in FMS at the aggregate level. Also, disturbances and integration effects are to be taken into account in APP of FMS.

5.7 Part-selection problems

A number of similarity measures and their related clustering or heuristic methods have been proposed (Kusiak [103,104]; Chakravarty and Shtub [36] in the literature for part-selection problems in FMS. However, these methods ignore the constraints that are present in the part selection problem. Furthermore, the similarity measures are often subjective, consisting of a mix of nominal and numerical data. Therefore, suitable quantitative aspects of similarity coefficients need to be established. The following situations which exist in practice are to be considered for investigation in part-selection problems;

1. a part has many attributes;

2. different part types may share common tools; and

3. optimal solution requires simultaneously partitioning the production order into the minimal number of batches.

Also, constraints such as tool magazine capacity, pallet size, speed and capacity of the machines and material handling system, etc., are to be considered while modeling the problem of part selection in FMS.

5.8   Resource grouping problems

In the majority of resource grouping problems, firms have used classification and coding schemes as tools in applying GT. However, if users identify that managerial and technical barriers are to be resolved in successfully applying GT, then significant and varied operational and strategic benefits could be achieved. Further, GT would be an integral and important part of future CAD/CAM activities at their plant. Establishing appropriate criteria for different types of resources grouping is very important. For example, the criteria for machine grouping may not be the same for part grouping problems. Multiple criteria would benefit GT applications to a greater extent. The GT principle can also be applied to storage systems, jigs and fixtures. Currently similarity coefficients typically rely on subjective measures such as common machines required in processing for evaluating the similarity between parts. Similarity coefficients that employ direct measures such as similarities between parts in terms of set-ups and tooling may yield good results. Another possible way to improve the cell formation procedures is through the use of part volume data, since it is a part of CAM, and its impact on group technology and the design of technology cells will be subject to further research.

Despite numerous economic benefits and operational advantages offered by the GT concept, its real potential has not been fully explored. A number of factors, including machine breakdown, under-utilization of resources and eventual unbalanced workload distribution in a multi-cell plant, present some problems when using the GT concept. These problems mostly come from standard principles of GT, such as the avoidance of interaction between the cells, and tendency to setting up permanent idealistic cells. Many managers are more comfortable with buying a piece of hardware than investing in (simulation) software. Therefore, a modular FMS simulator can be rather a cheap and useful tool in helping the user to select the appropriate operating rules for the system at hand. Only few firms have applied in practice the concept of GT to design, process planning (including NC programming), sales, purchasing, cost estimation, tooling, scheduling, new equipment selection, and tool selection due to the lack of literature available dealing with the application of GT in these problem areas (Wemmerlov and Hyer [206]).

5.9   Production ratio determination

Dynamic optimization models for generating flow control directives in FMC are very important. Therefore, models are to be developed for determining the production

26

ratio dynamically (on-line control), as day-to-day operation changes frequently owing to various reasons such as machine breakdowns, emergency order, etc. The important factors that are to be included in the production ratio determination are: (i) storage capacity, (ii) work-in-process inventory; (iii) material handling equipments capacity: (iv) the speed of the machines and AGVS; (v) pallet size; (vi) tool magazine capacity; etc. The reported models in the literature considered only few of these factors while estimating the production ratio. Therefore, all relevant factors are to be considered simultaneously while determining the production ratio for minimum throughput time and maximum productivity. In addition, the transportation time of parts has been neglected in the modeling of FMS determining the production ratio.

In the research area of FMS, much past work has been concentrated upon a unique system analyzing production control strategies. In some cases, no substantial system was available as a reference, necessitating a variety of system assumptions by researchers. It has been very difficult to compare control strategy effectiveness among applications due to a general lack of standardization among system definitions. Further, if the length of a control period is too large, control will be overly reactive to changes in the environment, thereby causing operational instability. These issues need further investigation.

5.10   Resource allocation and loading problems

The models that are reported in the literature cannot be implemented for large problems because of the 0-1 and general nature of the integer programming formulations. (Divakar et al. [50]). Hence, there is a need to develop heuristics or approximate procedures to solve these problems. Some of them can be:

1. suggest algorithms for flow control in FMS with dynamic set-up for a more realistic FMS;

2. integrate the concept of JIT in the part-selection problems in multi-stage FMS; and

3. develop a prototype simulation system for unmanned FMS.

Special attention should be given to non-exhaustive queue selection heuristics which may have great potential to perform well under a large variety of experimental conditions. Moreover, periodic routing-control problems associated with a variety of FMS such as non-homogeneous population systems, non-predetermined output proportion systems, and limited central buffer systems are to be considered for further investigation. In the FMS literature, tool management is simply ignored for modeling and analysis, although it influences the performance of loading and scheduling in FMS. Since grouping of parts and tools are inter-related to each other, an integrated grouping of parts and tools in FMS is to be considered for future research. The research objective is to obtain some general indications on modes of approaching loading and dispatching of FMS, taking into account that the performances of the rules are influenced by the configuration of the plant, and, on the other hand, that for a given plant, there is the further influence of the production mix.

Avonts et al. [13] suggested a simple linear programming model for determining the type of products and the quantities to be produced in a real FMS. They also demonstrated how one can select strategies using the LP model. Developing procedures for studying the effect of organizational constraints on cell sizes and product mix changes is also a challenging future research area. For the single part case, Kusiak [106] developed an approach, but for the multiple parts case, there is hardly any research available. This requires development of an integrated model for facility layout planning, process planning and the machine loading problem. Many Flexible Assembly Systems (FAS), which are normally composed of assembly robots and automated material handling equipment, exhibit considerable scope in the allocation of work load among machines. Transfer line production models can be applied in some of the cases. Further, the concept of critical resource can be used for resources like jigs and fixtures (Shankar and Srinivasulu [162]).

5.11   Measure of flexibility and performance

Researchers have defined various types of manufacturing flexibility and provided methods for measuring them. Nevertheless, opportunity costs associated with adding manufacturing flexibility such as set-up, part waiting, equipment idle, and inventory are one of the important manufacturing performance measures of FMS (Son and Park [174]). Factors such as disturbances, set-up times, loading and unloading time of parts were not considered in flexibility measures. Besides, labour flexibility is also to be taken into account while evaluating system performance. Part characteristics are not included while defining and evaluating various flexibilities. However, the latter influence the design of flexible manufacturing cells (Gupta and Goyal [169]).

Different types of flexibility can be unified in a structured system and a system's response to changes can be derived from components' flexibilities and other basic characteristics. The inter-relationship between various aspects of flexibility and other basic characteristics of production systems is complex and dependent on the expected production composition. Therefore, further research is needed for:

1. quantification of machine, routing, and control flexibilities;

2. development of models for the interconnections of these flexibilites in multi-machine systems; and

3. clarification of the relations between long-term and short-term flexibilities.

In real-life FMS, different types of parts are being processed simultaneously, therefore, such situations should be given due consideration while developing methods for measuring flexibility and its performance. A study to determine experimentally the influence of different kinds of flexibility on the performance parameters of an FMS under the aspect of different loading strategies and various system configurations may be an interesting research. Knowledge of the flexibility trade-offs can help the management to support the manufacturing strategy of the firm.

5.12 Artificial Intelligence and Expert Systems

Most of the Expert Systems developed are specifically applicable to scheduling problems for job shop, assembly line, and flow shop. Very few ES have been developed recently for solving machine layout problems in FMS. However, the literature that is dealing with knowledge-based Expert Systems for planning and scheduling in FMS using AI is rather limited. Furthermore, AI and ES are hardly exits for solving the problems of part selection, production ratio determination, equipment selection, and justification methods. Moreover, application of existing AI and ES can be enhanced by adding more decision-making capabilities to the existing knowledge bases. Realizing the importance of AI and ES in solving the complex problems of FMS on real-time control, there is a need to develop such systems. In addition, developing ES and AI for the following issues can be considered for future research: (i) layout problems, (ii) quality control, (iii) production ratio determination; and (iv) storage and robotic systems design. Moreover, Kusiak [112] and Kusiak and Chen [111] proposed a number of potential research problems to deal with the applications of AI and Expert Systems in FMS. In addition, there is no AI or Expert System available to deal with the problems of MRP, GT, and scheduling in FMS.

5.13 Additional future research directions

The following additional problems can be considered for future research:

1. Quality control issues in FMS, such as location of inspection stations, number of inspectors required and the type of equipment required for testing, are to be studied with appropriate models.

2. Safety considerations and environmental issues are to be addressed adequately in modeling the FMS environment.

3. The softwares required to operate FMS has not been given due attention in the literature. Therefore, models are to be developed for selecting suitable software for FMS, considering various operational and technological constraints.

4. Development of planning and scheduling methodologies for each class of FMS [108].

5. Study of the appropriateness of the presently used tools and techniques in FMS to design a hybrid approach.

6. Design of a feasibility study to evaluate economical and technical feasibilities of FMS.

7. Study of the implications of lot-sizing and Just-in-time concept in FMS.

8. Development of AI and Expert Systems for evaluating the flexibilities and economic benefits that result from flexibility.

6 Concluding remarks

In this paper, the literature available on FMS has been reviewed based on an appropriate classification scheme, in order to identify the gap between theory and practice, and future research problems. There are many models in FMS lacking integration while modeling the issues of various inter-related sub-systems in FMS. It would be more practical to develop models for the efficient operation of many sub-systems as in integrated FMS, for example, interaction between different manufacturing cells in an FMS is to be accounted for modeling and analysis. The scheduling of tool movement which is very important, but it received little attention from researchers and this can be considered for further studies and improvements. Artificial Intelligence and Expert Systems are to be developed for part-selection problems, resource grouping, material handling system design, and equipment selection. The constraints on capacity of the machine tools, tool magazine, number of pallets and buffer capacity, and machine breakdowns are to be incorporated while designing FMS. Simulation and pilot projects may be helpful to solve these problems. Further, the problems of operators/supervisors in FMS with respect to work design can be considered for further investigation. Safety and environmental aspects are to be incorporated in the decision making of FMS layouts. In addition, a suitable framework is to be established for studying the maintenance and quality and control aspects in FMS.

Acknowledgements

The authors are grateful to Professor Alan Mercer and two anonymous referees for their extremely useful and helpful comments on the earlier version of this manuscript. The authors also acknowledge the financial support by the Neste Foundation for this research work.

References

[1] Afentakis, P., Solomon, M.M., and Millen, R.A., "The part-type selection problem", in: K.E. Stecke and R. Suri (eds.), *Proceedings of the Third ORSA/TIMS Conference on Flexible Manufacturing Systems - Operations Research Models and Applications*, 1989, 141-146.

[2] Afentakis, P., Millen, R.A., and Solomon, M.M., "Dynamic layout strategies for flexible manufacturing systems", *International Journal of Production Research* 28, (1990) 311-323.

[3] Aggarwal, S.C., "MRP, JIT, OPT, FMS", *Harvard Business Review* 3, (1985) 8-16.

[4] Airey, J., and Young, C., "Economic justification, counting the strategic benefits", in: *Proceedings, 2nd International Conference on Flexible Manufacturing Systems*, IFS, London, 1983, 549-554.

[5] Akella, R., Maimon, O., and Gershwin, S.B., "Value function approximation via linear programming for FMS scheduling", *International Journal of Production Research*, 28 (1990) 1459-1470.

[6] Alberti, N., and Diega, N.L., "Cost efficiency: K.E. Stecke and R. Suri (eds.), An index of operational performance of flexible automated production environments", *Proceedings of the Third ORSA/TIMS Conference on Flexible Manufacturing Systems - Operations Research Models and Applications* 1989, 67-72.

[7] Al-Qattan, I., "Designing flexible manufacturing cells using a branch and bound method", *International Journal of Production Research* 28, (1990), 325-336.

[8] Ammons, J.C., Lofgren, C.B., and McGinns, L.F., "A large scale loading problem in flexible assembly", *Annals of Operations Research* 3 (1985) 319-328.

[9] Arbel, A., and Seidmann, A., "Performance evaluation of flexible manufacturing systems", *IEEE Transactions on Systems, Man, and Cybernetics* 14 (1984) 606-617.

[10] Ashayeri, J., and Gelders, L.F., "Interactive GPSS-PC Program Generator for automated material handling systems", *The International Journal of Advanced Manufacturing Technology* 2 (1987) 63-77.

[11] Askin, R.G., and Chiu, K.S., "A graph partitioning procedure for machine assignment and cell formation in group technology", *International Journal of Production Research* 28 (1990) 1555-1572.

[12] Avlonitis, G.J., and Parkinson, S.T., "The adoption of flexible manufacturing systems in British and German companies", *Industrial Marketing Management* 15 (1986) 97-108.

[13] Avonts, L.H., Gelders, L.F., and Van Wassenhove, L.N., "Allocation work between an FMS and a conventional jobshop: A case study", *European Journal of Operational Research* 33 (1988) 245-256.

[14] Azzone, G., and Bertele, O., "Measuring the economic effectiveness of flexible automation: A new approach", *International Journal of Production Research* 27 (1989) 735-746.

[15] Ballakur, A., and Steudel, H.I., "A within-cell utilization based heuristic for designing cellular manufacturing systems", *International Journal of Production Research* 25 (1987) 639-665.

[16] Barad, M., and Sipper, D., "Flexibility in manufacturing systems", *International Journal of Production Research* 26 (1988) 237-248.

[17] Berrada, M., and Stecke, H.E., "A branch and bound approach for FMS machine loading", in : *Proceedings of the First ORSA/TIMS Special Interest Conference on Flexible Manufacturing Systems: Operations Research and Applications*, University of Michigan, Ann Arbor MI, August 15-17, 1984.

[18] Blackburn, J., and Millen, R., "Perspectives of flexibility in manufacturing", in: A. Kusiak (ed.), *Modelling and Design of Flexible Manufacturing Systems*, Elsevier Science Publishers, Amsterdam, 1986, 157-170.

[19] Blazewicz, J., Sethi, S.P., and Sriskandarajah, C., "Scheduling of robot moves and parts in a robotic cells, in : K.E. Stecke and R. Suri (eds.), *Proceedings of the Third ORSA/TIMS Conference on Flexible Manufacturing Systems - Operations Research Models and Applications*, 1989, 133-138.

[20] Boctor, F.F., "Alternative formulations of the machine-part cell formation problem", in: K.E. Stecke and R. Suri (eds.), *Proceedings of the Third ORSA/TIMS Conference on Flexible Manufacturing Systems - Operations Research Models and Applications*, 1989, 133-139.

[21] Bozer, Y.A., and Srinivasan, M.M., "Tandem configuration for automated guided vehicle systems offer simplicity and flexibility", *Industrial Engineering* 21 (1989) 23-27.

[22] Brill, P.H., and Mandelbaum, M., "Measures of flexibility for production systems", in : *Proceedings of the IXth International Conference on Production Research*, 1987, 2472-2481.

[23] Brill, P.H., and Mandelbaum, M., "On measures of flexibility in manufacturing systems", *International Journal of Production Research* 27 (1989) 747-756.

[24] Browne, J., Dobois, D., Rathmill, K., Sethi, S.P., and Stecke, K., "Classification of flexible manufacturing systems", *The FMS Magazine* 2 (1984) 114-117.

[25] Burbidge, J.L., *Introduction to Group Technology*, Wiley, New York, 1975.

[26] Burstein, M.C., and Talbi, M., "Economic justification for the introduction of flexible manufacturing technology: Traditional procedures versus a dynamic-based approach", in: Proceedings of the First ORSA/TIMS Conference on FMS, Ann Arbor, MI, August 1984.

[27] Buzacott, J.A., and Shanthikumar, I.G., "Models for understanding flexible manufacturing systems", *American Institution of Industrial Engineers Transactions* 12 (1980) 339-349.

[28] Buzacott, J.A., "The fundamental principles of flexibility in manufacturing systems", in : *Proceedings of the 1st International Conference on Flexible Manufacturing Systems*, Brighton, UK 1982, 13-22.

[29] Buzacott, J.A., and Mandeibaum, M., "Flexibility and productivity in manufacturing systems", in: *Proceedings of the IIE Conference*, Chicago, 1985, 404-413.

[30] Buzacott, J.A., and Yao, D.D., "Flexible manufacturing systems: A review of analytical models", *Management Science* 31 (1986) 890-905.

[31] Cariapa, V., "Multimode machine tools - A concept that improves operations of flexible manufacturing systems", *International Journal of Production Research* 29 (1991) 1069-1079.

[32] Carlsson, B., "Flexibility and the theory of the firm", *International Journal of Industrial Organization* 7 (1989).

[33] Carrie, A.S., and Perera, D.T.S., "Work scheduling in FMS under tool availability constraint", *International Journal of Production Research* 24 (1986) 1299-1308.

[34] Cassandras, C.G., "Dynamic material handling for a class of flexible manufacturing systems", *Annals of Operations Research* 3 (1985) 427-448.

[35] Cassandras, C.G., "Autonomous material handling in computer integrated manufacturing systems", in : A. Kusiak (ed.), *Modelling and Design of flexible manufacturing systems*, Elsevier Science Publishers, Amsterdam, 1989, 81-89.

[36] Chakravarty, A.K., and Shtub, A., "Selecting parts and loading flexible manufacturing systems", in : *Proceedings of the first ORSA/TIMS Special Interest Conference on Flexible Manufacturing Systems*, University of Michigan, Ann Arbor, MI, 1984, 284-289.

[37] Chakravarty, A.K., and Shtub, A., "Production planning with flexibilities in capacity", in : *Proceedings of the Second ORSA/TIMS Special Interest Conference on Flexible Manufacturing Systems*, University of Michigan, Ann Arbor, MI, 1986, 333-343.

[38] Chakravarty, A.K., and Shtub, A., "Capacity, cost and scheduling analysis for a multiproduct flexible manufacturing cell", *International Journal of Production Research* 25 (1987) 1143-1156.

[39] Chan, H.M., and Miller, D.A., "Direct clustering algorithm for group formation in cellular manufacture", *Journal of Manufacturing Systems* 1 (1982) 64-76.

[40] Chandrasekharan, M.P., and Rajagopalan, R., "An ideal-seed non-hierarchical clustering algorithm for cellular manufacturing", *International Journal of Production Research* 24 (1986) 451-454.

[41] Chandrasekharan, M.P., and Rajagopalan, R., "ZODIAC: An algorithm for concurrent formation of part families and machine cells", *International Journal of Production Research* 25 (1987) 835-850.

[42] Chang, Y.L., Matsuo, H., and Sullivan, R.S., "A bottleneck-based beam search for job scheduling in a flexible manufacturing system", *International Journal of Production Research* 27 (1989) 1949-1961.

[43] Chen, I., and Chung, C.H., "Effects of loading, and routeing decisions on performance of flexible manufacturing systems", *International Journal of Production Research* 29 (1991) 2209-2225.

[44] Chryssolouris, G., Pierce, J.E., and Cobb, W., "A decision-making approach to the operation of FMS", in: K.E. Stecke and R. Suri (eds.), *Proceedings of the Third ORS/TIMS Conference on Flexible Manufacturing Systems - Operations Research Models and Applications*, 1989, 133-139.

[45] Chu, C.H., and Tsai, M., "A comparison of three array-based clustering techniques for manufacturing cell formation", *International Journal of Production Research* 28 (1990) 1417-1433.

[46] Chung, C.H., and Chen, I.J., "A systematic assessment of the value of flexibility for an FMS", in : K.E. Stecke and R. Suri (eds.). *Proceedings of the Third ORSA/TIMS Conference on Flexible Manufacturing Systems - Operations Research Models and Applications*, 1989, 27-34.

[47] Chung, S.H., and Lee, T.R., "A heuristic method for solving FMS master production scheduling problem", In: K.E. Stecke and R. Suri (eds.), *Proceedings of the Third ORSA/TIMS Conference on Flexible Manufacturing Systems - Operations Research Models and Applications*, 1989, 127-132.

[48] Dirne, C.W.G., "Planning problems of flexible automated manufacturing cells in a job shop - A case study", in : K.E. Stecke and R. Suri (eds.), *Proceedings of the Third ORSA/TIMS Conference on Flexible Manufacturing Systems - Operations Research Models and Applications*, 1989, 61-66.

[49] Dirne,C.W.G., "The quasi-simultaneous finishing of work orders on a flexible automated manufacturing cell in a job shop", *International Journal of Production Research* 28 (1990) 1635-1655.

[50] Divakar, R., Singh, N., and Aneja, Y.P., "Integrated design of cellular manufacturing systems in the presence of alternative process plans", *International Journal of Production Research*, 28 (1990) 1541-1554.

[51] Doulgeri, Z., Hibberd, R.D., and Hushand, T.M., "The scheduling of flexible manufacturing systems", *Annals of the CIRP* 36 (1987) 1-14.

[52] Dunkler, O., Mitchell, C.M., Govindaraj, T., and Ammons, J.C., "The effectiveness of supervisor control strategies in scheduling flexible manufacturing systems", *IEEE Transactions on Systems, Man, and Cybernetics* 18, (1988) 233-237.

[53] Dutta, S.P., Lashkari, R.S., Nadoli, G., and Ravi, T., "A heuristic procedure for determining manufacturing families from design based grouping for FMS", *Computers and Industrial Engineering* 10 (1986) 193-201.

[54] Edghill, J.S., and Davies, A., "Flexible manufacturing systems - The myth and reality", *International Journal of Advanced Manufacturing Technology*, 1 (1985) 37-54.

[55] Egbelu, P.J., and Tanchoco, J.M.A., "Characterization of automatic guided vehicle dispatching rules", *International Journal of Production Research* 22 (1984) 359-374.

[56] Egbleu, P.J., and Tanchoco, J.M.A., "Potentials for bidirectional guide path for automated guided vehicle based systems", *International Journal of Production Research* 24, (1986), 1075-1097.

[57] Elsayed, E.A., and Kao, T.Y., "Machine assignments in production systems with manufacturing cells", *International Journal of Production Research* 28 (1990), 489-501.

[58] El-Tamimi, A.M., Suliman, S.M.A., and Williams, D.F., "A simulation study of part sequencing in a flexible assembly cell", *International Journal of Production Research* 27 (1989) 1769-1793.

34

[59] Eversheim, W., and Hermann, P., "Recent trends in flexible automated manufacturing", *Journal of Manufacturing Systems* 1 (1982) 139-147.

[60] Fazakerly, G., "Research report on the human aspects of group technology and cellular manufacturing", *International Journal of Production Research* 14 (1976) 123-135.

[61] Garetti, M., Pozzetti, A., and Bareggi, A., "On-line loading and dispatching in flexible manufacturing systems", *International Journal of Production Research* 28 (1990) 1271-1292.

[62] Gaskins, R.J., and Tanchoco, I.M.A., "Flow path design for automated guided vehicle system", *International Journal of Production Research* 25 (1987) 667-676.

[63] Gelders, L.F., and Ashayeri, J., "Management issues in flexible process manufacturing", in: C.F.H. van Rijn (ed.), *Logistics - Where Ends Have To Meet*, Pergamon Press, Amsterdam, 1987.

[64] Ghosh, B.K., "Equipment investment decision analysis in cellular manufacturing", *International Journal of Operations & Production Management* 10 (1990) 5-20.

[65] Greene, T.J., and Sadowski, R.P., "Cellular manufacturing control", *Journal of Manufacturing Systems* 2 (1983) 137-145.

[66] Gross, J., "Intelligent feedback control for flexible manufacturing systems", Ph.D Thesis, University of Illinois at Urbana Champaign, IL, 1987.

[67] Graham, M., and Rosenthal, S., "Flexible manufacturing systems require flexible people", *Human Systems Management* 6 (1986) 211-222.

[68] Gupta, Y.P., "Human aspects of flexible manufacturing systems", *Production and Inventory Management* (1989) 30-36.

[69] Gupta, Y.P., and Goyal, A., "Flexibility of manufacturing systems: Concepts and measurements", *European Journal of Operational Research* 43 (1989) 119-135.

[70] Gupta, T., and Seifoddini, H., "Production data based similarity coefficient for machine-component grouping decisions in the design of a cellular manufacturing system", *International Journal of Production Research* 28 (1990) 1247-1269.

[71] Gustavsson, S., "Flexibility and productivity in complex production processes", *International Journal of Production Research* 22 (1984) 801-808.

[72] Han, M.H., and McGinnis, L.F., "Flow control in flexible manufacturing: Minimization of stockout cost", *International Journal of Production Research* 27 (1989) 701-715.

[73] Han, M.H., Na., Y.K., and Hogg, G.L., "Real-time tool control and job dispatching in flexible manufacturing systems", *International Journal of Production Research* 27 (1989) 1257-1267.

[74] Harmonosky, C.M., and Sadowski, R.P., "The system generator concept for FMS research and analysis", *International Journal of Production Research* 28 (1990) 559-571.

[75] Heragu, S.S., and Kusiak, A., "Machine layout problem in flexible manufacturing systems", *Operations Research* 32 (1988) 258-268.

[76] Heragu, S.S., and Kusiak, A., "Machine layout: An optimization and knowledge-based approach", *International Journal of Production Research* 28 (1990) 615-635.

[77] Hildebrand, R.R., "Scheduling flexible machine systems when machines are prone to failures", Ph.D. Thesis, MIT, Cambridge, MA, 1982.

[78] Hintz, G.W., and Zimmermann, H.J., "A method to control Flexible Manufacturing Systems", *European Journal of Operational Research* 41 (1989) 321-334.

[79] Hitomi, R.R., Yoshimura, M., and Higashimoto, A., "Design and effective use of automatic setup equipment for flexible manufacturing cells", in: *Proceedings of Japan-U.S.A. Symposium on Flexible Automation*, 1986, 559-566.

[80] Hitomi, K. Yoshimura, M., and Ohashi, K., "Design and scheduling for flexible manufacturing cells with automatic set-up equipment", *International Journal of Production Research* 27 (1989) 1137-1147.

[81] Hough, P.G., "Cost estimating for the factory of the future", in: K.E. Stecke and R. Suri (eds.), *Proceedings of the Third ORSA/TIMS Conference on Flexible Manufacturing Systems - Operations Research models and Applications*, 1989, 53-60.

[82] Howell, R., and Stephen, R.S., "Capital investment in the new manufacturing environment", *Management Accounting* LXIX (1987) 26-32.

[83] Huber, V., and Hyer, N.L., "The human impact of cellular manufacturing", *Journal of Operations Management* 4 (1985) 213-228.

[84] Hutchinson, G.K., "Flexibility is key to economic feasibility of automating small batch manufacturing, *Industrial Engineering* 21 (1984) 77-86.

[85] Hwang, S.S., and Shogan, A.W., "Modeling and solving an FMS part selection problem", *International Journal of Production Research* 27 (1989) 1349-1366.

[86] Hwang, S.S., "Models for production planning in flexible manufacturing systems", Ph.D Dissertation, University of California, Berkeley, CA, 1986.

[87] Hwang, S.S., and Shanthikumar, J.G., "An FMS production planning system and evaluation of part selection approaches", Management Science Working Paper No. MS-43, School of Business Administration, University of California, Berkeley, CA 1987.

[88] Hyer N.L., and Wemmerlöv, U., "MRP/GT: A framework for production planning and control of cellular manufacturing", *Decision Sciences* 13 (1982) 681-701.

[89] Jaikumar, R., Flexible manufacturing systems: Management perspective", WP 1-784-078, Division of Research, Harvard Business School, 1984.

[90] Jaikumar, R., "Postindustrial manufacturing", *Harvard Business Review* 64 (1986) 69-76.

[91] Kalkunte, M., Sarin, S.C., and Wilhelm, w.E., "Flexible manufacturing systems: A review of modelling approaches for design, justification and operation", In: A. Kusiak (ed.), *Modelling and Design of Flexible Manufacturing Systems*, Elsevier Science Publishers, Amsterdam, 1986, 3-25.

[92] Kimemia, J.G., "Hierarchical control of production in flexible manufacturing systems", Ph.D. Thesis, LIDS-TH-1215, MIT, Cambridge, MA, 1982.

[93] Kimemia, J.G., and Gershwin, S.B., "An algorithm for computer control of a flexible manufacturing system", *IIE Transactions* 15 (1983) 353-362.

[94] Kiran, A.S., and Tansel, B.C., "Optimal pickup point problem on material handling network", *International Journal of Production Research* 27 (1989) 342-356.

[95] Kouvelis, P., and Kiran, A.S., "The plant layout problem in automated manufacturing systems", Working Paper, ISE Department, USC, 1988.

[96] Kouvelis, P., and Kiran, A.S., "Layout problem in flexile manufacturing systems: Recent research results and further research directions", in: K.E. Stecke and R. Suri (eds.), *Proceedings of the Third ORSA/TIMS Conference on Flexible Manufacturing Systems - Operations Research Models and Applications*, 1989, 147-152.

[97] Kouvelis, P., and Lee, H.L., "The material handling design of automated manufacturing systems: A graph theoretic modelling framework", Working Paper, University of Texas at Austin, Austin, TX, 1988.

[98] Krajeski, L., King, B., and Ritzman, L.P., "Kanban MRP and shaping the manufacturing environment", *Management Science* 33 (1987) 39-57.

[99] Kumar, K.R., Kusiak, A., and Vanneli, A., "Grouping of parts and components in FMSs", *European Journal of Operational Research* 24 (1986) 387-400.

[100] Kumar, P., Singh, N., and Tewari, N.K., "A nonlinear goal programming model for the loading problem in a flexible manufacturing system", *International Journal of Production Research* 2 (1987) 13-20.

[101] Kumar, P., Tewari, N.K., and Singh, N., "Joint consideration of grouping and loading problems in a flexible manufacturing system" *International Journal of Production Research* 28 (1990) 1345-1356.

[102] Kumar, V., "Entropy measures of measuring manufacturing flexibility", *International Journal of Production Research* 25 (1987) 957-966.

[103] Kusiak, A., "Part families selection model for flexible manufacturing systems", in: *American Industrial Engineering Conference Proceedings*, 1983, 575-580.

[104] Kusiak, A., "The part families problem in flexible manufacturing systems", in: *Proceedings of the First ORSA/TIMS Conference on Flexible Manufacturing Systems*, University of Michigan Ann Arbor, MI, 1984, 237-242.

[105] Kusiak, A., "Flexible manufacturing systems: A structural approach", *International Journal of Production Research* 23 (1985) 1057-1073.

[106] Kusiak, A., "The part families problem in FMS", *Annals of Operations Research* 3 (1985) 117-124.

[107] Kusiak, A., *Modelling and Design of Flexible Manufacturing Systems*, North-Holland Amsterdam, 1986.

[108] Kusiak, A., "Application of Operational Research models and techniques in flexible manufacturing systems", *European Journal of Operational Research* 24 (1986) 336-345.

[109] Kusiak, A., "Artificial Intelligence and operations research in flexible manufacturing systems", *Information Systems and Operations Research* 25 (1987) 2-12.

[110] Kusiak, A., "The production equipment requirements problem", *International Journal of Production Research* 25 (1987) 319-325.

[111] Kusiak, A., and Chen, M., "Expert systems for planning and scheduling manufacturing systems", *European Journal of Operational Research* 34 (1988) 113-130.

[112] Kusiak, A., and Heragu, S., "Knowledge based system for machine layout", in: *Proceedings, International Industrial Engineering Conference*, Institute of Industrial Engineers, Orlando, FL, 1988, 159-167.

[113] Kusiak, A., "Knowledge engineering and optimization in automated manufacturing systems", in: *Proceedings of the 2nd International Conference on Production Systems*, IN-RIA, April 6-10, 1987, Paris, France.

[114] Kusiak, A., *Intelligent Manufacturing Systems*, Prentice Hall, Englewood Cliffs, NJ, 1989.

[115] Kynast, R., "Flexibility in high production manufacturing", in: *Flexible High Production Machining Systems*, Society of Manufacturing Engineers, Dearborn, MI, 1987.

[116] Lashkari, R.S., Dutta, S.P., and Padhye, A.M., "A new formulation of operation allocation problem in flexible manufacturing systems: Mathematical modelling and computational experience", *International Journal of Production Research* 25 (1987) 1267-1283.

[117] Lee, H.F., Srinivasan, M.M., and Yano, C.A., "An algorithm for the minimum cost configuration problem in flexible manufacturing systems", in: K.E. Stecje and R. Suri (eds.) *Proceedings of the Third ORSA/TIMS Conference on Flexible Manufacturing Systems - Operations Research Models and Applications*, 1989, 85-92.

[118] Levary, R.R., and Kalehik, S., "Facilities layout - survey of solution procedures", *Computers and Industrial Engineering* 9 (1985) 141-153.

[119] Liang, M., and Dutta, S.P., "A mixed integer programming approach to the machine loading and process planning problem in a process layout environment", *International Journal of Production Research* 28 (1990) 1471-1484.

[120] Mahadevan, B., and Narendran, T.T., "Design of an automated guided vehicle based material handling system for a flexible manufacturing system", *International Journal of Production Research* 28 (1990) 1611-1622.

38

[121] Maley, J., Ruiz-Myer, S., and Solberg, J., "Dynamic control in automated manufacturing: A knowledge integrated approach", *International Journal of Production Research* 26 (1988) 1739-1748.

[122] Mandelbaum, M., and Buzacott, J.A., "Flexibility and its use: A formal decision process and manufacturing view", in: *Proceedings of the Second ORSA/TIMS Special International Conference on Flexible Manufacturing Systems*, Ann Arbor, MI, 1986, 119-130.

[123] Mandelbaum, M., and Buzaccott, J., "Flexibility and decision making", *European Journal of Operational Research* 46 (1990) 263-273.

[124] Mandelbaum, M., and Brill, P.H., "Examples of measurement of flexibility and adaptivity in manufacturing systems", *Journal of the Operational Research Society* 40 (1989) 603-609.

[125] Maxwell, W.L., and Muckstadt, J.A., "Design of automatic guided vehicle systems", *IIE Transactions* 14 (1982) 114-124.

[126] Mazzola, J.B., "Heuristics for the FMS/MRP rough-cut capacity planning problem", in: K.E. Stecke and R. Suri (eds.), *Proceedings of the Third ORSA/TIMS Conference on Flexible Manufacturing Systems - Operations Research Models and Applications*, 1989, 119-126.

[127] Mazzola, J.B., Neebe, A.W., and Dunn, C.V.R., "Production planning of a flexible manufacturing system in a material requirement planning environment", *The International Journal of Flexible Manufacturing Systems* 1 (1989) 115-142.

[128] Meredith, J.R., and Suresh, N.C., "Justification techniques for advanced manufacturing technologies", *International Journal of Production Research* 24 (1986) 1043-1057.

[129] Meredith, J. and Marianne, H., "Justifying new manufacturing systems: A managerial approach", *Solan Management Review* 28 (1987) 49-62.

[130] Mills, D.E., "Flexibility as a manufacturing objective", *International Journal of Industrial Organization* 4 (1986) 302-315.

[131] Monahan, G.E., and Smunt, T.L., "The flexible manufacturing investment decision", Working Paper, Graduate School of Business Administration, Washington University, St. Louis, 1984.

[132] Montazeri, M., Gelders, L.F., and Van Wassenhove, L.N., "A modular simulator for design, planning and control of flexible manufacturing systems", *The International Journal of Advanced Manufacturing Technology* 3 (1988) 15-32.

[133] Montazeri, M., and Van Wassenhove, L.N., "Analysis of scheduling rules for an FMS", *International Journal of Production Research* 28 (1990) 785-802.

[134] Nof, S.Y., Barash, M.M., and Solberg, J.J., "Operational control of item flow in versatile manufacturing systems", *International Journal of Production Research* 12, (1979) 479-493.

[135] Nof, S.Y., Whinston, A.B., and Bullers, W.I., "Control and decision support in versatile manufacturing systems", *American Institution of Industrial Engineers* 12 (1980) 156-165.

[136] O'Grady, P.J., and Menon, U., "A multiple criterion approach for production planning of automated manufacturing", *Engineering Optimization* 8 (1985) 161-175.

[137] O'Grady, P.J., and Menon, U., "Loading a flexible manufacturing system", *International Journal of Production Research* 25 (1987) 1053-1068.

[138] O'Grady, P.J., and Lee, K., "An intelligent cell control system for automated manufacturing", *International Journal of Production Research* 26 (1988) 845-862.

[139] Ozden, M., "A simulation study of multiple-load-carrying automated guided vehicles in a flexible manufacturing system", *International Journal of Production Research* 26 (1988) 1353-1366.

[140] Park, C.S., and Son, Y.K., "An economic evaluation model for advanced manufacturing systems", *The Engineering Economist* 34 (1988) 1-26.

[141] Park. S., Raman, N., and Shaw, M.J., "Heuristic learning for pattern-directed scheduling in a flexible manufacturing system", in: K.E. Stecke and R. Suri (eds.), *Proceedings of the Third ORSA/TIMS Conference on Flexible Manufacturing Systems - Operations Research Models and Applications* 1989, 133-139.

[142] Pollard, P.L., and Tapscott, Jr., R.L., "Investment evaluation methodology", in: K.E. Stecke and R. Suri (eds.), *Proceedings of the Third ORSA/TIMS Conference on Flexible Manufacturing Systems - Operations Research Models and Applications*, 1989, 47-52.

[143] Primrose, P.L., and Leonard, R., "Establish the viability of FMS"., *The FMS Magazine* 3 (1985) 114-116.

[144] Purcheck, G., "Machine-component group formation: An heuristic method for flexible production cells and flexible manufacturing systems", *International Journal of Production Research* 23 (1985) 911-943.

[145] Rabelo, L.C., and Alptekin, S., "Synergy of neural networks and expert systems: A knowledge based approach", in: K.E. Stecke and R. Suri (eds.), *Proceedings of the Third ORSA/TIMS Conference on Flexible Manufacturing Systems - Operations Research Models and Applications*, 1989, 133-139.

[146] Rajagopalan, R., and Batra, J.L., "Design of cellular production systems - a graph theoretic approach", *International Journal of Production Research* 13 (1975) 567-579.

[147] Raja, Gunasingh, K. and Lashkari, R.S., "Machine grouping problem in cellular manufacturing systems - an integer programming approach", *International Journal of Production Research* 27 (1989) 1465-1473.

[148] Raman, N., Talbot, B., and Rachamadugu, R., "Simultaneous scheduling of material handling devices in automated manufacturing", in : K.E. Stecke and R. Suri (eds.), *Proceedings of the Second ORSA/TIMS Conference on FMS: Operational Research Models and Applications*, Elsevier Science Publishers, Amsterdam, 1986.

[149] Ro, I.K., and Kim, J.I., "Multi-criteria operational control rules in flexible manufacturing systems (FMS)", *International Journal of Production Research* 28 (1990) 47-63.

[150] Sabuncouglu, I., and Hommertzheim, D.L., "An investigation of machine and AGV scheduling rules in an FMS", in: K.E. Stecke and R. Suri (eds.), *Proceedings of the Third ORSA/TIMS Conference on Flexible Manufacturing Systems - Operations Research Models and Applications*, 1989, 79-84.

[151] Sarin, S.C., "Computerized analysis of robotized production cell", *Information Systems and Operations Research* 25 (1987) 46-56.

[152] Sarin, S.C., and Chen, S.C., "The machine loading and tool allocation problem in a flexible manufacturing system", *International Journal of Production Research* 25 (1987) 1081-1094.

[153] Sarkar, B.R., "Optimum manpower models for a production system with varying production rates", *European Journal of Operational Research* 24 (1986) 147-154.

[154] Sawik, T., "Modelling and scheduling of a flexible manufacturing system", *European Journal of Operational Research* 45 (1990) 177-190.

[155] Schweitzer, P.J., and Seidmann, A., "Processing rate optimization for FMS's with distinct multiple job visits to work centers", in: K.E. Stecke and R. Suri (eds.), *Proceedings of the Third ORSA/TIMS Conference on Flexible Manufacturing Systems - Operations Research Models and Applications*, 1989, 79-84.

[156] Seidmann, A., and Nof. S.Y., "Unitary manufacturing cell design with random product feedback flow", *IIE transactions* 17 (1985) 188-193.

[157] Seidmann, A., and Schweitzer, P.J., "Part selection policy for a flexible manufacturing cell feeding several production lines", *IIE Transactions* 16 (1984) 355-362.

[158] Seidmann, A., Schweitzer, P.J., and Nof., S.Y., "Performance evaluation of a manufacturing cell with random multiproduct feedback flow", *International Journal of Production Research* 23 (1985) 1171-1184.

[159] Shafer, S.M., and Meredith, J.R., "A comparison of selected manufacturing cell formation techniques", *International Journal of Production Research* 28 (1990) 661-673.

[160] Shalev-Oren, S., Seidmann, A., and Schweitzer, P.J., "Analysis of flexible manufacturing systems with priority scheduling: PMVA", *Annals of Operations Research* 6 (1985) 115-139.

[161] Shankar, K., and Tzen, Y.J., J., "A loading and dispatching problem in a random flexible manufacturing system", *International Journal of Production Research* 23, (1985) 579-595.

[162] Shankar, K., and Srinivasalu, A., "Some solution methodologies for loading problems in a flexible manufacturing system", *International Journal of Production Research* 27 (1989) 1019-1034.

[163] Shanthikumar, J.G., and Stecke, K.E., "Reducing working-process inventory in certain classes of flexible manufacturing systems", *European Journal of Operational Research* 26 (1986) 266-271.

[164] Sharifnia, A., Caramanis, M., and Gershwin, S.B., "Dynamic set-up scheduling and flow control in flexible manufacturing systems", in: K.E. Stecke and R. Suri (eds.), *Proceedings of the Third ORSA/TIMS Conference on Flexible Manufacturing Systems - Operations Research Models and Applications*, 1989, 133-139.

[165] Sharit, J., "Supervisory control of a flexible manufacturing system", *Human Factors* 27 (1985) 47-59.

[166] Sharit, J., and Elhence, S., "Computerization of tool-replacement decision making in flexible manufacturing systems: A human-systems perspective", *International Journal of Production Research* 27 (1989) 2027-2039.

[167] Shaw, M.J., "A pattern-directed approach to FMS scheduling", *Annals of Operations Research* 15 (1988) 353-376.

[168] Shewchuk, I., "Justifying flexible automation", *American Machinist* October (1984) 93-96.

[169] Shtub, A., "Modelling group technology cell formation as a generalised assignment problem", *International Journal of Production Research* 27 (1989) 775-782.

[170] Slack, N., "Flexibility as a manufacturing objective", *International Journal of Operations & Production Management* 3 (1983) 4-13.

[171] Solberg, J.J., and Nof, S.Y., "Analysis of material flow control in alternative manufacturing configurations", *Journal of Dynamic Systems, Measurement and Control* 1980.

[172] Solot, P., "A concept for planning and scheduling in an FMS", *European Journal of Operational Research* 45 (1990) 85-95.

[173] Son, Y.K., and Park, C.S., "Economic measure of productivity, quality and flexibility in advanced manufacturing systems", *Journal of Manufacturing Systems* 6 (1987) 193-207.

[174] Son, Y.K., and Park, C.S., "Quantifying opportunity costs associated with adding manufacturnig flexibility", *International Journal of Production Research* 28 (1990) 1183-1194.

[175] Spur, G., and Mertins, K., "Flexible manufacturing systems in Germany, conditions and development trends", in: *FMS: Proceedings of the first conference, IFS*, Kempston, Bedforshire, 1982.

[176] Stecke, K.E., and Solberg, J.J., "The optimal planning of computerized manufacturing systems - CMS loading problem", Report 20, NSF Grant APR 74 15256, School of Industrial Engineering, Purdue University, West Lafayette, IN, 1979.

[177] Stecke, K.E., "Production planning problems for flexible manufacturing systems", Ph.D. Dissertation, Purdue University, West Lafayette, IN, 1981.

[178] Stecke, K.E. and Solberg, J.J., Loading and control polcies for a flexible manufacturing system", *International Journal of Production Research* 19 (1981) 481-490.

42

[179] Stecke, K., "Formulation and solution of non-linear integer production planning problem for flexible manufacturing systems", *Management Science* 29 (1983) 272-285.

[180] Stecke, K.E., "Design planning scheduling and control problems of flexible manufacturing systems", *Annals of Operations Research* 3 (1985) 3-12.

[181] Stecke, K.E., and Solberg, J.J., "The optimality of the unbalanced workloads and machine group sized for FMS", *Operations Research* 33 (1985) 882-910.

[182] Stecke, K., and Talbot, B., "Heuristic loading algorithms for flexible manufacturing systems", Working Paper No. 348, Division of Research, Graduate School of Business Administration, The University of Michigan, Ann Arbor, MI, 1986.

[183] Stecke, K.E., "A hierarchical approach to solving machine grouping and loading problems of flexible manufacturing systems", *European Journal of Operational Research* 24 (1986) 369-378.

[184] Steudel, H.J., and Ballakur, A., "A dynamic programming based heuristic for machine grouping in manufacturing cell formation", *Computers and Industrial Engineering* 12 (1987) 215-222.

[185] Suresh, N.C., "Towards an integrated evaluation of flexible automation investments", *International Journal of Production Research* 28 (1990) 1657-1672.

[186] Suresh, N.C., and Sarkis, J., "An MIP formulation for the phased implementation of FMS modules", in K.E. Stecke and R. Suri (eds.) *Proceedings of the Third ORSA/TIMS Conference on Flexible Manufacturing Systems - Operations Research Models and Applications* 1989, 41-46.

[187] Suri, R., and Hilderbrant, R.R., "Modeling flexible manufacturing systems using mean-value analysis", *Journal of Manufacturing Systems* 3 (1984) 27-38.

[188] Suri, R., and Whitney, C.K., "Decision support requirements in flexible manufacturing", *Journal of Manufacturing Systems* 3 (1984) 61-69.

[189] Swamidass, P.M., "Manufacturing flexibility: Strategic issues", Discussion Paper No. 305, Graduate School of Business, Indiana University, 1988.

[190] Tang, C.S., "A job scheduling model for a flexible manufacturing machine", in: *1986 IEEE International Conference on Robotica and Automation*, 1 (1986) 152-155.

[191] Tansel, B.T., and Kiran, A.S., Optimum storage location in Flexible Manufacturing Cells", *Journal of Manufacturing Systems* 7 (1988) 121-129.

[192] Tatikonda, M.V., and Crosheck, M.K., "A case study on FMS capacity determination", in: K.E. Stecke and R. Suri (eds.), *Proceedings of the Third ORSA/TIMS Conference on Flexible Manufacturing Systems - Operations Research models and Applications*, 1989, 73-78.

[193] Taymaz, E., "Types of flexibility in a single-machine production system", *International Journal of Production Research* 27 (1989) 1891-1899.

[194] Tempelmeier, H., Kuhn H and Tetzlaff, U., "Performance evaluation of flexible manufacturing systems with blocking", *International Journal of Production Research* 27 (1989) 1963-1979.

[195] Thomasma, T., and Hilbrecht, K., "Specification methods for material handling control algorithms in flexible manufacturing systems", in: K.E. Stecke and R. Suri (eds.), *Proceedings of the Third ORSA/TIMS Conference on Flexible Manufacturing Systems - Operations Research Models and Applications*, 1989, 79-84.

[196] Tomkins, J.A., and Reed, R., "An applied model for the facilities design problem", *International Journal of Production Research* 14 (1976) 583-595.

[197] Trunk, C., "Simulation for success in the automatic factory", *Material Handling Engineering* 44 (1989) 64-76.

[198] Van Looveren, A.J., Gelders, L.F., and Van Wassenhove, L.N., "A review of FMS planning models", in : A. Kusiak (ed.), *Modelling and Design of Flexible Manufacturing Systems*, Elsevier Science Publishers, Amsterdam, 1986, 3-31.

[199] Vanneli, A., and Kumar, K.R., "A method for finding minimal bottleneck cells for grouping part-machine families", *International Journal of Production Research* 24 (1986) 387-399.

[200] Ventura, J.A., Chen, F.F., and Wu, C.H., "Grouping parts and tools in flexible manufacturing systems production planning", *International Journal of Production Research* 28 (1990) 1039-1056.

[201] Vinod, A.B., and Solberg, J.J., "The optimal design of flexible manufacturing systems", *International Journal of Production Research* 23 (1985) 1141-1151.

[202] Waghodekar, P.H., and Sahu, S., "Machine-component cell formation in GT: MACE", *International Journal of Production Research* 25 (1987) 937-948.

[203] Wallace, W.J., and Thuesen, G.I., "Annotated bibliography of investing in flexible automation", *The Engineering Economist* 32 (1987) 247-257.

[204] Wemmerlöv, U and Hyer, N.L., "Comments on direct clustering algorithm for group formation in cellular manufacture", *Journal of Manufacturing Systems* 3 (1984) 7-9.

[205] Wemmerlöv, U. and Hyer, N.L., "Procedures for the part family/machine group identification problem in cellular manufacture", *Journal of Operations Management* 6 (1986) 125-147.

[206] Wemmerlöv, U. and Hyer, N.L., "Research issues in cellular manufacturing", *International Journal of Production Research* 25 (1987) 413-431.

[207] Wemmerlov, U., *Production Planning and Control Procedures for Cellular Manufacturing Systems: Concepts and Practice*, American Production and Inventory Control Society, Falls Church, VA 1988.

[208] Wemmoerlöv, U. and Hyer, N.L., "Cellular manufacturing in the US industry: A survey of users", *International Journal of Production Research* 27 (1989) 1511-1530.

44

[209] Whitney, C.K., and Suri, R., "Algorithms for part and machine selection in flexible manufacturing systems", *Annals of Operations Research* 3 (1985) 34-45.

[210] Widmer, M., and Solot, P., "Do not forget the breakdowns and the maintenance operations in FMS design problems", *International Journal of Producgtion Research* 28 (1990) 421-430.

[211] Wilhelm, W.E., and Shin, H.Y., "Effectiveness of alternate operations in a flexible manufacturing system", *International Journal of Production Research* 23 (1985) 65-79.

[212] Wilson, J.M., "An alternative formulation of the operation-allocation problem in flexible manufacturing systems", *International Journal of Production Research* 27 (1989) 1405-1412.

[213] Wu, H.L., Venugopal, R., and Barash, M.M., "Design of a cellular manufacturing system: A syntactic pattern recognition approach", *Journal of Manufacturing Systems* 5 (1986) 81-88.

[214] Wu, S.Y.D., and Wysk, R.A., "Multi-pass expert control system - A control/scheduling structure for flexible manufacturing cells", *Journal of Manufacturing Systems* 7 (1988) 107-120.

[215] Wu, S.Y.D., and Wysk, R.A., "an application of discrete-event simulation to on-line control and scheduling in flexible manufacturing", *International Journal of Production Research* 27 (1989) 1603-1623.

[216] Yao, D.D., "Flexibility - A condition for effective production systems", *Material Flow* 2 (1985) 143-149.

[217] Yao, D.D., and Buzacott, J.A., "Modelling a class of state-dependent routeing in flexible manufacturing systems", *Annals of Operations Research* 3 (1985) 153-167.

[218] Yao, D.D., and Buzacott, J.A., "Models of flexible manufacturing systems with limited local buffers", *International Journal of Production Research* 24 (1986) 107-118.

[219] Yih, Y. and Thesen, A., "Semi-Markov decision models for real-time scheduling", *International Journal of Production Research* 29 (1991) 2331-2346.

Flexible Manufacturing Systems: Recent Developments
A. Raouf and M. Ben-Daya (Editors)

FMS-Planning Since the Early Eighties - Experiences, Recommendations and Future Tendencies

H.-J. Warnecke[a], R. Steinhilper[b] and H. Storn[b]

[a]President of Fraunhofer-Gesellschaft, Leonrodstr. 54, 80636 München, Germany

[b]Fraunhofer Institute for Manufacturing Engineering and Automation (IPA),
Nobelstr. 12, 70569 Stuttgart, Germany

This paper gives an outline of contents, main reasons, planning and developing procedures, success determining factors, results and also side effects of nearly 100 FMS-related projects carried out for industrial companies by the department "Manufacturing Systems" at the Fraunhofer-Institute for Manufacturing Engineering and Automation (IPA) during the period 1979 until 1991. Furthermore goals and resulting requirements concerning manufacturing systems of the Nineties are discussed with an emphasis on future control systems and demands for personnel.

1. FMS-PLANNING OF THE EIGHTIES REVEALED DEVELOPMENT GAPS IN AUTOMATION EQUIPMENT.

The Fraunhofer Institute for Manufacturing Engineering and Automation (IPA), which is linked to the Institute of Industrial Production and Corporate Management (IFF) of Stuttgart University, has been involved in the planning of Flexible Manufacturing Systems (FMS) for more than a dozen years now, doing FMS-work mainly for industrial customers /1/, /2/. Since 1979 until 1991 almost 100 FMS-related projects have been carried out, while most projects contained planning tasks concerning the machining and material flow system. Although many FMS components have been commercially available in the late Eighties, so that one could expect the main task of an FMS project to consist of the accurate planning and arrangement of known components, there had to be noticed a steadily significant amount of new development tasks for FMS, first of all for new machining and material flow equipment (Figure 1).

46

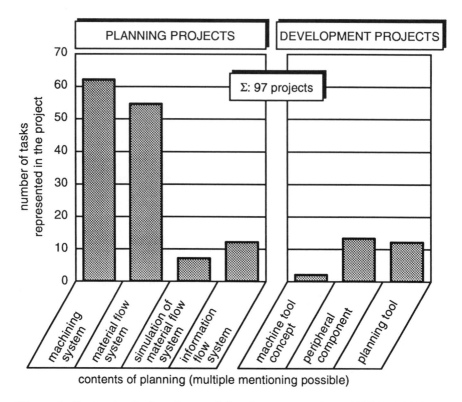

Figure 1. Contents of planning and development projects (1979 - 1990)

One reason for this may be the fact that the products to be manufactured in FMS have changed: whereas FMS for drilling, milling and boring have been among the "pioneering" systems, today also FMS for turning, sheet metal working as well as welding and even for wooden or plastic parts machining are designed and developed (Figure 2). So consequently the most important side effect experienced in IPA's customer's companies during 70 projects aiming on the planning and installation of a Flexible Manufacturing System or Cell has been the successful development of a new automation device (Figure 3).

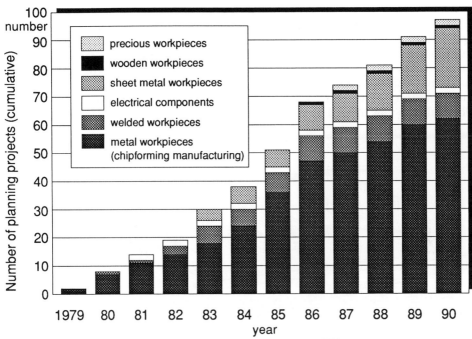

Figure 2. Types of workpieces at FMS-planning since 1979

side-effect	number of cases
☐ ideas for other planning contents	7
☐ ideas for new products	6
☐ improved product design	16
☐ improved operating sequences	14
☐ other organizational benefits	7
☐ ideas for developing new automation equipment	21*

* among them: 15 successfully realized

Figure 3. Side-effects caused by FMS-planning in 76 planning-projects

An important result of FMS-related development work during the past years arose from the nuisance of lacking pallet systems being compatible not only to one certain manufacturer's machine tools or manufacturing cells but to several, in order to being able to combine different brands and even types of machine tools within one FMS: the modular and standardized workpiece pallet for rotational parts, which has jointly been developed by IPA and more than ten manufacturers of machine tools and automation equipment (Figure 4 and 5). This pallet is now offered by various suppliers and has come into frequent use.

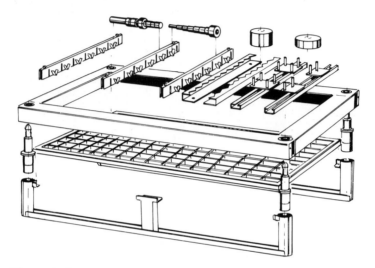

Figure 4. Standardized modular pallet (elements)

Another example for new developed automation equipment is a programmable unloading device, applicable to sheet metal manufacturing cells, the necessity of which had been realized during an examination on "Manufacturing Systems for complex thin-walled parts".

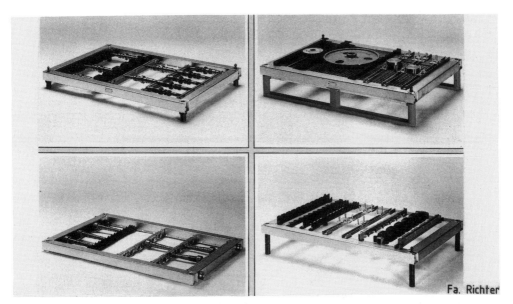

Figure 5. Standardized modular pallets (applications)

2. THE NEW ROLE OF FMS IS: INTEGRATION BUT AUTOMATION.

Prognoses of the expectable changes of the factory automation market prove a share of up to 50 % for software to be the most important portion of factory automation equipment and FMS of the Nineties (Figure 6). The increasing importance of information flow planning and development within FMS projects is also shown in Figure 7: It can be expected that in a few years their ratio will reach nearly one half of the project contents.

Due to the above mentioned market changes and to grown ratings of information flow planning, today's planning procedures (Figure 8) must be reviewed in order to cut consecutive costs. As most mistakes that are discovered during application of an FMS preferably emerged during planning (Figure 9), the integration of all planning contents, in particular the information flow system, is a "must". In other words: The planning of the machining system, material flow system and information flow system has to be started on from the beginning of an FMS-project in order to achieve an overall harmonization of FMS-planning (Figure 10).

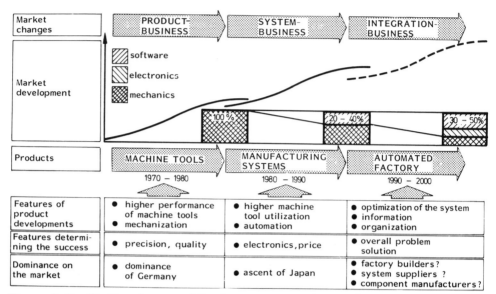

Figure 6. Qualitative changes of the factory automation market until 2000 according to Mc Kinsey

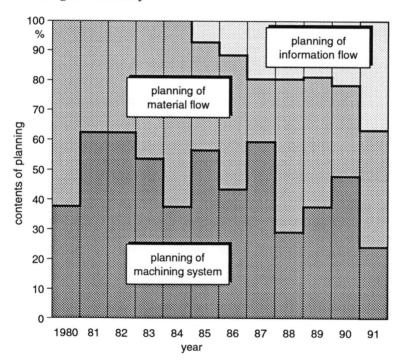

Figure 7. Development of contents of FMS-planning since 1980

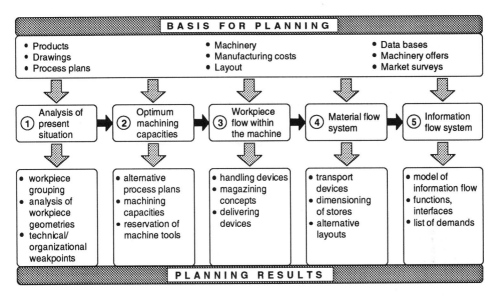

Figure 8. Recent procedure for FMS-planning

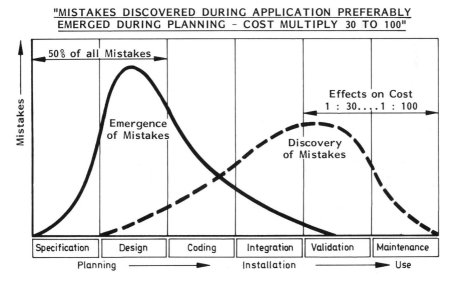

Figure 9. Emergence and discovery of mistakes

52

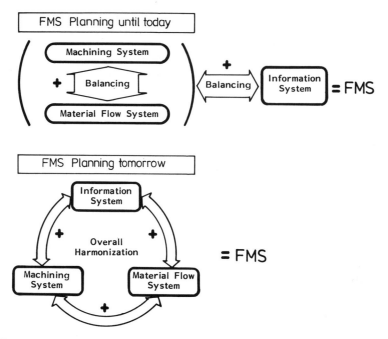

Figure 10. Harmonization of FMS-planning

3. THE MOST SUCCESS-DETERMINING FACTOR DURING FMS-PLANNING REMAINS THE SAME: ATTITUDE.

The increase of the importance of development tasks during FMS-planning and the necessity of considering information flow concepts for FMS-operation also affected the average planning effort and duration. While the planning effort, e.g. the number of engineers involved, has increased considerably and doubled since 1979 - from 2 to 3 engineers on IPA side and from 2 to 5 engineers on customer's side - however, no extension of the average project duration was allowed by the customers in order to keep up with their competitors (Figure 11).

Because Flexible Manufacturing Systems are in many cases considered as a shop floor based start-up for Computer-Integrated-Manufacturing (CIM) - and therefore members of all company departments contribute to and benefit from the work of the FMS-project team - the planning and introduction of FMS needs and means a new way of thinking in many areas throughout the company. Knowing this, the most important success-determining factors during FMS planning and development cannot be described by technological data and performances. Experiences of IPA's FMS-planning during the last decade show that the key for success mainly lies in the attitude of the people involved in planning, design, installation and operation of FMS. Only in those cases, where the project team's members completely identified with the task, where teamwork and the ability for cooperation and coordination in all phases of the project and between all participants characterized the introduction of the system, successfully operating equipment and satisfactory figures have been the result.

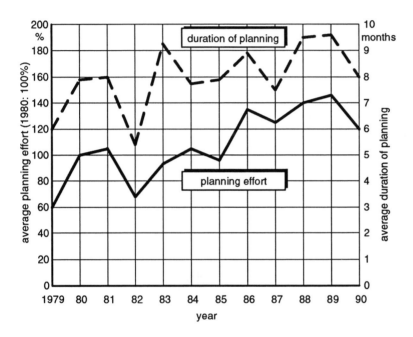

Figure 11. Planning effort and duration of planning at FMS-projects since 1979

Even a dozen years after the appearance of the first systems, FMS can still not be regarded as turn key solutions, but still need a lot of pioneering spirit and the readiness to face many changes that will occur not only in the manufacturing shop but in many departments and at all stages of order handling. Especially what information flow is concerned, those project teams who intensively looked for an adequate supplier of controls and software had much less success than those who agreed to contribute by themselves to the new developments and modules necessary to run a complex system carefully tailored to the particular needs of their company. They were prepared to acquire a big amount of knowledge and knowhow, which cannot just be bought from a supplier, but will emerge during planning, installation and operation of the system. Of course carefulness in selection of the right suppliers of machining, material flow and

54

information flow equipment as well as security and accuracy at all stages of the planning procedure are just as important as teamwork and joint responsibility.

Especially the latter decisively affects the correctness of the data input such as production batch sizes, machining and quality requirements and operation sequences of the workpieces that determine the design and the dimensioning of the system. Whereas conceptional planning and detailed planning of the system can be done mainly by the consulting institute, this data input and its accuracy will always be the task and remain to the responsibility of the customer (Figure 12).

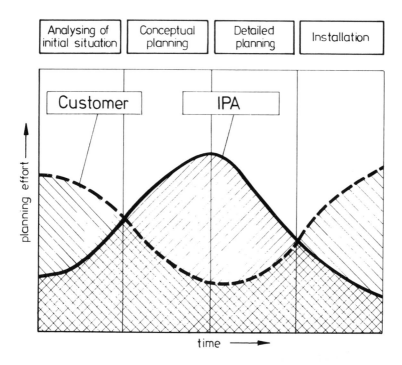

Figure 12. Planning efforts of project-partners at FMS-planning

4. PLANNING AND DEVELOPMENT RESULTS - EXAMPLES OF SUCCESSFUL INSTALLATIONS

70 of nearly 100 FMS-related projects mentioned in the previous chapters aimed at the planning and installation of a Flexible Manufacturing System or Cell or the development of a new solution for machining, material flow or information flow. More than one half of these projects led to the realization of the original object (Figure 13). In many other cases at least some parts of the originally planned installation could be realized - often as a first step towards an FMS which would be installed in the long run.

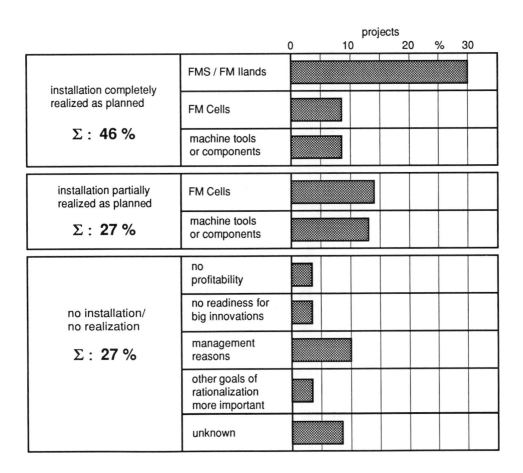

Figure 13. Percentage of installations at FMS-planning-projects

56

Figures 14 to 16 show two examples among the most outstanding joint realizations of Flexible Manufacturing Systems during the past five years. In the FMS shown in Figure 14 precision tools are manufactured, representing the shop floor based approach of a medium sized company to Computer Integrated Manufacturing (CIM). It is in operation at Montanwerke Walter Company in Tübingen, West Germany and includes five axis machining, fully computerized control of material flow, both for workpieces and tools, and a modular computer hierarchy with connection to CAD and PPC. Figures 15 and 16 show the layout of and a view into an FMS for automotive front-wheel-drive-parts, which is installed at Birfield Trasmissioni in Brunico, Italy. It is characterized by three shift operation of more than a dozen Flexible Manufacturing Cells connected by wire guided carriers, responsible for the transport of workpiece pallet containers, thus representing one of the most advanced Flexible Manufacturing Systems for rotational parts worldwide.

The key for success of these systems has been the fact that the user company took a lot of efforts to develop components as well as control software jointly with IPA or by themselves instead of expecting turn key solutions from suppliers.

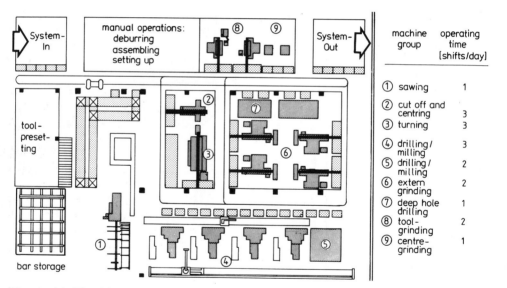

Figure 14. Flexible Manufacturing System for precision parts

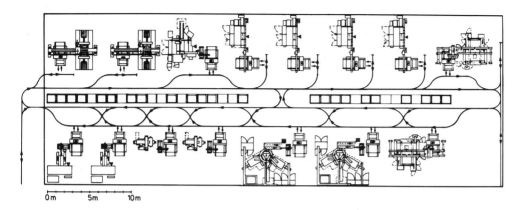

Figure 15. FMS for rotational parts, (soft machining area)

Figure 16. FMS for rotational parts (storage)

5. FMS OF THE NINETIES: THEIR SHAPES ARE DETERMINED BY EFFECTIVE ORGANIZATION.

Besides considerable development challenges of the Nineties concerning the design of machine tools and their periphery, the consideration of the information flow system of an FMS at an early stage of planning will have to be emphasized in order to minimize consecutive costs, planning failures could have been responsible for. This decision may be of even greater importance, the larger an FMS grows and the bigger organizational efforts have to be to run the system. But FMS or other computerized systems of the Nineties need not necessarily look big, if integration aspects are taken into account, that keep the organizational efforts at a low level.

Many users complain the huge share of waiting times of the workpieces between the various individual operations such as turning, milling, grinding etc. especially within small and medium batch manufacturing. In practice one setup often means one week of throughput time, thus easily leading to a two or three month duration of an order in the manufacturing shop.

These facts counteract all efforts to fasten deliveries. In many cases this problem cannot be solved by an interlinking of different machine tools to an automated system, because each work-piece differs from the other so that cycle times would not correspond sufficiently. Also the handling and interlinking devices would not be flexible enough to cover all flexibility requirements of the part spectrum, which would lead to even more setup work, not only for the machine tools but also for peripheral equipment and, important enough, for manufacturing control. One way to keep these efforts as low as possible is the integration of functions already at the machine tool level.

At present, many developments concerning machine tools point in the direction of integrating different manufacturing processes in one machine tool. Originally started within lathes as well as machining centres, turning, 5-face-drilling and -milling, grinding and even 3-dimensional measuring can now be performed in one setup (Figure 17).

An important gap still remains concerning heat treatment, which uses to make at least two different setups between the machining of "soft" and hardened workpieces inevitable. The appearance of first applications of inductive heat treatment devices on machining centres (Figure 18) however gives hope, that in the near future the horizon of completely automated machining of a workpiece in one setup comes nearer.

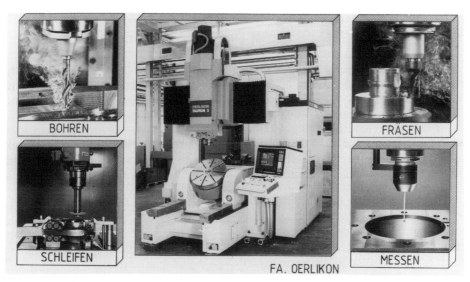

Figure 17. CNC-machining centre for finishing of precision parts

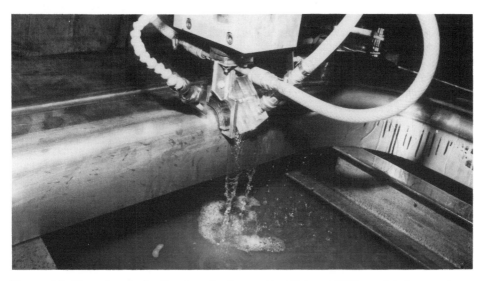

Figure 18. Use of an inductive hardening unit within a CNC-machining centre

60

6. THE KEY TO REASONABLE FMS-FLEXIBILITY IS A MODULAR DESIGN OF CONTROL SOFTWARE.

As technical tendencies of machine tool controls and of Flexible Manufacturing Systems show a large variety of goals and resulting requirements concerning FMS control (Figure 19) demands for new solutions. In order to provide a "learning-by-doing" possibility in a step-by-step approach from stand-alone computer controlled automation islands to integrated systems with a multi-level computer hierarchy, a modular design of software packages will be absolutely necessary. These packages should be able to be assembled in different combinations according to different individual applications and needs.

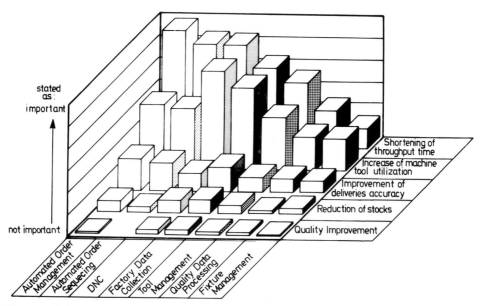

Figure 19. Main goals and resulting requirements to FMS-control

Furthermore, FMS controls and software have to be compatible with business information systems in a way they can successfully interact with the overall system using a common database for design, manufacturing and all other tasks within the entire operation, thus becoming a step towards and a part of computer integrated manufacturing (CIM) in the automated factory of the future.

Since there is no standard factory, effective integration concepts can only be created on the basis of exact knowledge of the production sequences, the operational sequences and the necessary information architecture of the participating departments. To facilitate this overall concept, some planning work will initially be necessary, since CIM is primarily concerned with planning and organization.

The technical obstacles are buried in the lack of compatibility between computers and software. Interface standards are, in this case and in cases mentioned above, the key elements for the future spread of CIM applications. It will certainly take some time, however, until all those involved in decision-making processes have overcome their reservation with regard to CIM. In this respect it should be noted that time critical processes all pose greater problems than mere MAP. Lacking coordination between

- existing system software
- bought-in application software from different sources and
- in-house software

obstructs integration and productivity of the service for electronic data processing. The shortage of personnel will further increase the demand on system software suppliers with respect to training and support. Also lacking is modular software which can be distributed with little adaptation for various applications -and thus more cheaply for the individual. 50 % "standard", the "remainder" adaptation - these are rather optimistic hopes and estimates. Practice generally shows that a tailor-made solution must be developed from scratch and also that it will never be complete. CIM involves enough changes within a company so that, as a rule, the software should be adapted to the organization of the company and not vice versa. This means, however, that even the mere specification of a complex solution offered by a potential supplier will have to be paid for.

7. FMS OF THE FUTURE NEED FUTURE PRODUCTION PLANNING AND CONTROL SYSTEMS.

The way a company performs production planning and control (PPC) belongs to its strategy for competitiveness. There are two fundamental competitive strategies: Mass production enables production costs to be reduced. Customization however puts less emphasis on costs. The strategy adopted affects all areas of the company's activities, e.g. quality assurance and production control (Figure 20). No existing company conforms totally to a particular one of these two widely differing models of strategy. In reality a company lies somewhere in between these extremes. As a result there is no such thing like an ideal organizational structure. Each company is exposed to different operational conditions. This applies equally to production planning and control.

Production planning and control systems currently in use solely employ a single strategy in decision situations such as batch formation and capacity calculations. Consequently no attempt is made to optimize production control. Often, it is not even obvious that there is a decision situation. The justification in using such systems is found merely on the premise that experts in the field could not come up with a better solution in the same time.

62

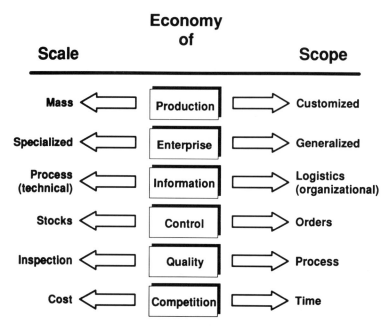

Figure 20. Organizational characteristics of mass- and short-run production

However, assuming that suitable industrial hardware was available for collecting data, thus any information on the process being planned/controlled would be available without delay, a future production control system might be regarded as an "optimum system" if

a) the planning horizon is filled with tasks/orders,
b) out of all possible alternative strategies the best solution is chosen by means of appropriate evaluation,
c) whenever the master data are altered, the entire planning horizon is suitably updated in accordance with a) and b).

A system of this type will always have some form of limitation, even assuming there was unlimited computer power and capacity, since the following conditions would have to be met:

- Unlimited amount of time would have to be available for updating the master data and for executing planning operations.
- The production process would have to be interrupted in order to keep the master data up to date for the time needed to carry out the calculations. Otherwise the effects of control would be inaccurate from the start.

Production control systems avoid this problem by only processing new production plans when production is stopped for an extended period of time, such as at night or at weekends, and not at the time when changes might be needed (i.e. for planning requirements).

If the requirements for up-to-date data are higher and remain constant, as is the case of CIM concepts, then computation must be carried out during production. To attain more accurate and up-to-date data without interrupting production, other system concepts are needed. A production control system of this type must carry out four distinct functions:

1. Assess error messages to establish whether reevaluation is necessary or whether the deviation is tolerable.
2. Prepare the data required for the work process to continue on the basis of the assessment of the production situation, such as information on the next order/job.
3. Temporary memorization of the next event when information has to be passed to the production process (e.g. commencement of the next order).
4. Updating of the planning horizon up to completion, taking into account that the next event must receive current data.

This concept is depicted in Figure 21.

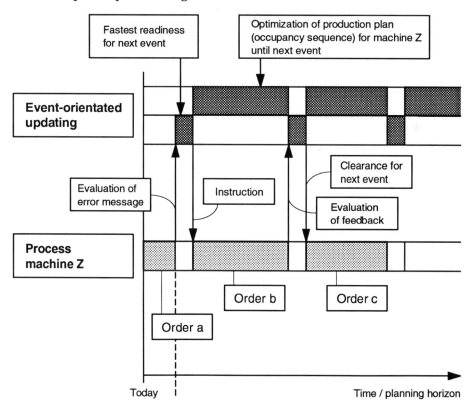

Figure 21. Functions of a future PPC-system: Feedback from production process

8. KEEP IN MIND THE HUMAN ASPECT: TODAY'S INVESTMENTS IN PERSONNEL DETERMINE TOMORROW'S PAYBACKS.

Besides modern hard- and software, people have become more important in automated manufacturing systems than anybody expected a few years ago when the unmanned factory seemed to come into use within a few years. Today we know that the latter event will not happen necessarily in the middle term. In spite of that there is a big demand on qualified men and women for control and other operations, as in most cases of "scrap" in FMS human errors are the main reasons (Figure 22).

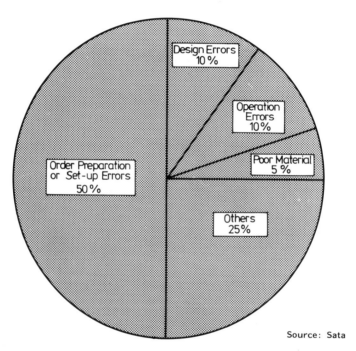

Figure 22. Reasons for "scrap" in Flexible Manufacturing Systems

As more and more different tasks and capabilities are integrated within one setup or one machine tool we will also have to integrate more and more different knowledge and qualification within those persons working with FMS, which leads to considerable changes in job contents and a reduction of work division (Figure 23 and 24). The workers have to know the mechanical problems of the parts as well as they need the know how of programming the machine tools and the capability of acting as a system operator. Therefore operations productivity and training have to be balanced. It is recognized that especially

- decision making capability
- cooperation and coordination of operations and
- knowledge

are increasingly required and force us to invest not only in hard- and software
but also in human capital when going towards FMS.

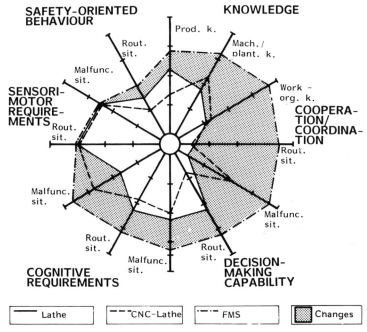

Figure 23. Changes in job contents

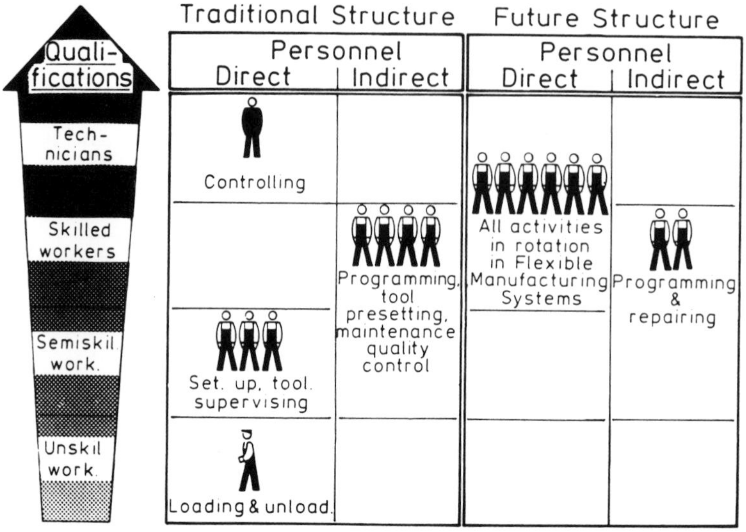

Figure 24. Reduction of work division

REFERENCES

/1/R. STEINHILPER, H. STORN and TH. REINHARD (1988) A Message from FMS-Work during 10 Years, Proc. 7th Int. Conf. on Flexible Manufacturing Systems, pp3-12, IFS Stuttgart

/2/H.-J. WARNECKE, R. STEINHILPER, H. STORN (1992) A Message from 100 FMS-Projects for Industry, Proc. 29th Int. MATADOR Conf., Univ. of Manchester

Part II

Flexibility Issues

Flexible Manufacturing Systems: Recent Developments
A. Raouf and M. Ben-Daya (Editors)
© 1995 Elsevier Science B.V. All rights reserved.

Manufacturing Systems: Flexibility Assessment

A. Raouf and M. Farooq Anjum

Systems Engineering Department, King Fahd University of Petroleum and Minerals,
Dhahran 31261, Saudi Arabia.

Abstract
The literature on Flexible manufacturing systems (FMS) does not provide an overall measure of flexibility for these systems. In this paper, we classify and discuss the different measure of flexibility presented in the literature and propose an overall measure.

1 Introduction

With the onset of the industrial revolution it was generally believed that it would be more profitable to manufacture products in terms of batches rather than producing them individually. While this would result in staggering the costs involved over the entire batch it would also lead to more work in progress inventory (WIP). At around the same time emphasis came to be laid upon the replacement of manual labour by machines. This was the process of mechanization. Automation followed whereby not only the individual processes or groups of processes were mechanized but also the associated systems such as material handling, control etc. This further on lead to the development of dedicated systems. While the dedicated systems are highly productive, but are least flexible.

Many different factors affect the strategies adopted by a firm. Some of these are the rapid rate of technological change, greater complexity of the manufacturing system itself, larger financial commitment with the accompanying increase in risk involved, increasing market expectations, and increased competition. All this requires that for a manufacturing firm to survive in today's markets it should be highly efficient.

Thus this points to a growing need for manufacturing systems that are both highly productive in order to meet the large demands of the marketplace and flexible in order to overcome the various uncertainties regarding the volume of the product, the attributes of the product etc. Hence in order to meet the conflicting requirements of high volume and flexibility, flexible manufacturing systems came into existence. It has been pointed out in Dixon(1992) [1] that the focus of competition in the global marketplace is shifting from quality and service towards flexibility.

There are quite many advantages to be obtained from incorporating flexibility into the manufacturing system. The advantages which can be achieved internally range from incorporation of a variety of operations, monitoring the tool life and tool performance, more emphasis on safety to small inventories, constant staffing level and less lead time between order and delivery. With respect to the external environment the advantages of flexibility include provision for off-line work and tool setting, easy assimilation of new products as well as little manning during operation. Thus a system possessing high degree of flexibility can achieve fast throughput, a variety of transformations as well as overcome any production snags.

Flexibility does not seem to have a universally accepted definition. The most commonly accepted definition of flexibility is the ability to take up different positions or alternatively the ability to adopt a range of states (Slack 1983 [2]). Flexibility is an indication of potential and hence is difficult to measure. Many different authors have defined many different types of flexibilities in the literature. In this paper our endeavour has been to investigate the different important measures of flexibility as given in the literature. We classify the different flexibility measures into short term and long term flexibilities. Then we look in detail at each measure of flexibility separately. Inconsistencies between the definitions of the different authors are also pointed out.

One aspect in which the literature on FMS seems to be inadequate is in letting the managers of a firm decide about the level of flexibility present in their systems. No integrated methods to measure the flexibility in a system are provided. Hence in this paper our intention is also to provide a method to measure flexibility in any given system.

The organisation of the paper is as follows. In Section 2 we look at the different definitions of elemental flexibilities which are important for the system. In Section 3 we present a method to measure overall flexibility of any given system. Finally Section 4 concludes this work.

2   Flexibility Definitions

In this section flexibility definitions in terms of various segments of manufacturing as developed by various authors are presented.

2.1   Machine flexibility

Browne et al.(1984) [3] defined machine flexibility as the ease of change to process a given set of part types. A measure can be obtained by computing the ratio of setup time to processing time. Buzacott(1982) [4] defines machine flexibility as the ability of the system to cope with changes and disturbances at the machines and workstations. Thus this is actually an indicator of the internal change within the system. Das and Nagendra (1993) [5] define this with respect to a workcentre as the ability of the machine or workcentre to perform more than one type of processing operation efficiently. Barad and Sipper(1988) [6] propose machine set up flexibility as an equivalent and more suggestive term for describing flexibility. This is because many of the activities contributing to this flexibility can be regarded as set up activities.

Carter(1986) [7] proposed a method to measure machine flexibility. He proposed to measure it by

1. number of tasks that can be performed by the machine.

2. range of possible dimensions.

3. cost incurred in making the changeovers.

4. time required to changeover to a different operations.

2.2   Process flexibility

Browne et al.(1984) [3] defined Process flexibility as the ability to produce a given set of part types. Mandelbaum(1978) has also called this as the action flexibility which relates to situations where decisions are made sequentially without knowledge of the future.

Son and Park (1987) [8] define this as the adaptability to various changes in part processing, such as in equipment and tools breakdown, random access of product mix, process schedule and so forth. Process flexibility also indicates a foregone opportunity to add value to materials processed. Poor process flexibility (large waiting cost or W.I.P) is a primary source of various manufacturing problems like poor quality. Sethi and Sethi(1990) [9] define this as the ability of the manufacturing system to produce a set of part types without major setups, which is also called as mix flexibility by Carter(1986). This is useful in reducing batch sizes and, in turn, inventory costs. Further the need to duplicate machines is also lessened as it is possible to share the machines.

Taymaz(1989) [10] has concentrated on aspects of flexibility specific to single machine systems. Processing at any time a mix of different parts which are loosely related to each other in some way gives a measure of the mix flexibility. In this way redundancy can be minimized by allowing machines to be shared among different products.

This has also been variously termed as job flexibility(Buzacott 1982 [4]) or design flexibility (Gerwin 1982 [11]) and Frazelle (1986) [12]. In fact there seems to be a difference between the definitions of Buzacott and Browne et al. in this respect. Buzacott doesn't differentiate between process variety within machine level and process variety which may be achieved through making use of different machines, whereas Browne et al. focus entirely on the product variety within machine level. As pointed out by Barad and Sipper(1988) [6] the process flexibility as defined by these authors actually represents a combination of process variety, interchangeability and redundancy of unit operations coupled with transfer possibilities, allowing utilization of the process variety for a given product mix. They further propose to decompose this flexibility into two potential flexibilities namely process flexibility which is defined as the system process variety, unrelated to a specific product mix and the transfer flexibility which is the system capability to move parts between machining centres. Even the transfer flexibility is unrelated to a specific product mix. This has also been called as the part specific flexibility by Chatterjee et al (1984) [13].

2.3   Product mix flexibility

Dixon(1992) [1] defined mix flexibility as the ability to manufacture a variety of products within a short period of time and without major modification of existing facilities. Gerwin(1983) [11] also gave a similar definition with the difference that he didn't specify a time frame. Mix flexibility is referred to as product-mix flexibility in Slack(1987)[2].

Job flexibility is defined by Buzacott(1982) [4] as the ability of the system to cope with changes in the jobs to be processed by the sytem. Thus this reflects upon the ability of the system to cope with external change.

A need for process/mix flexibility is created because of the uncertainty as to which products will be demanded by the customers. It has also been observed that this requires a low degree of specialisation for workers and equipment.

2.4 Routing Flexibility

Browne et al (1984) [3] define this as the ability to process a given set of parts on alternative machines. Taymaz(1989) [10] identifies that routing flexibility is associated with the dynamic assignment of machines. Uncertainty with respect to the machine downtime requires that a system possess routing flexibility.

According to Sethi and Sethi [9] this refers to the ability of the manufacturing system to produce a part by alternate routes through the system. Bernardo and Zubair (1992) [14] define routing flexibility as the ability of the system to continue producing a given part mix despite internal and/or external disturbances. Since routing flexibility can result from either the assigned routes that a part type actually uses or the potential routes possible, the authors propose two measures of routing flexibility – actual routing flexibility and potential routing flexibility. Actual routing flexibility is a measure of the number of existing production routes that could be used. Potential routing flexibility is a measure of the possible alternate routes of making the part.

It can be deduced that routing flexibility is product mix dependent. It is a measure of the means available for transferring the parts and also the ability to perform equivalent operations on a part in different locations. This facilitates production under conditions of breakdowns or other bottlenecks that may arise during the system operation. In fact this is a means of achieving mix flexibility.

Azzone and Bertele(1987) [15] narrow down this flexibility as to the ability of the system to operate with one or more machines not working. Kusiak(1986) [16] has called this as the scheduling flexibility. This can be because this flexibility affects the amount of freedom during scheduling phase.

2.5 Volume flexibility

Browne et al (1984) [3] define this as the ability to operate profitably at varying overall levels. The need for this flexibility is because of the uncertainty with regard to the amount of customer demand. Bernardo and Zubair(1992) in fact limit the definition by considering a single part. Graves(1988) [17] in fact calls volume flexibility as the rate flexibility which is given as the ratio of slack in production capacity normally available to the variability in the demand process.

This has also been called as the system set up flexibility by Barad and Sipper(1988)[6]. This is because lack of volume flexibility implies that low volumes wouldn't be able to economically justify the investment in system setup. Volume flexibility measures the ability to handle shifts in volume for a given part. Thus this allows the factory to adjust production within a wide range.

A system which has a workforce possessing varied skills and adjustable capacity requirements is likely to be more volume flexible. Further such a system would be capable of overcoming the uncertainty regarding the amount of customer demand for products.

2.6 Operation flexibility

Browne et al (1984) [3] define this as the ability to interchange ordering of operations on a part. Kumar (1986) [18] has also given a similar definition. This has also been called as the process sequence flexibility in Gupta and Goyal(1989)[19].

Sethi and Sethi(1990) also define operation flexibility similarly as the ability of a part to be produced in different ways. Thus a system possessing operation flexibility has the property of ease of scheduling of parts in real time.

But at the same time one is constrained by the design restrictions. The control costs can be the other major factor which prevent the random ordering of several operations on each part. Use of the modelling tool 'Petrinets' for the purpose of evaluating the operational flexibility of the system was done by Barad and Sipper(1988).

2.7 Customizing flexibility

Bernardo and Zubair (1992) [14] have defined this as the ability to process different mixes of parts on different flexible manufacturing systems in the same company.

2.8 Adaptation Flexibility

Zelenovic (1982) [20] defines the flexibility of a production system as the measure of its capacity to adapt to changing environmental conditions and process requirements. He also defines another dimension of flexibility called the adaptation flexibility as the value of time needed for system transformation/adaptation from one to another job task. It can be seen that this definition is quite similar to that of Browne et al. as given for process flexibility.

2.9 Equipment Flexibility

Son and Park(1987) [8] have defined Equipment flexibility as the capacity of equipment to accomodate new products and some variants of the existing products. They measure equipment flexibility in terms of idle cost (opportunity cost for equipment underutilization).

2.10 Path flexibility

Browne et al (1984) [3] define this as the possibility of having more than one path from the origin to the destination.

2.11 Program flexibility

Browne et al. (1984) [3] define this as the ability to operate under central computer control. But Sethi and Sethi (1990) differ in the definition of the program flexibility by bringing in the time factor in that they require the system to run unattended for a long enough period. Thus this improves the degree of automation of the system.

2.12 Capacity flexibility

Bernardo and Zubair(1992) [14] have given a measure of capacity flexibility which is the ability of the system to respond to unanticipated demand for a given part. It is mathematically described as the ratio of the number of extra units of a given part made from all routes to the average demand of that part.

2.13 Material Handling Flexibility

Sethi and Sethi(1990) [9] define material handling flexibility as the ability of the material handling systems to move different part types effectively through the manufacturing facility, including loading and unloading of parts, inter-machine transportation and storage of parts under various conditions of the manufacturing facility.

2.14 Delivery Flexibility

This has been defined in Slack and Correa (1992)[21] as the ability to adjust and meet adjusted delivery dates. In fact two "dimensions" of flexibility for product, mix, volume and delivery were also suggested by the authors. The two were range flexibility which was a measure of how much the system can change and response flexibility to measure how fast the system can change. Thus the resulting classes of flexibility were an indication of not only the rate of change which the system could cope with but also the level of change itself that the system could withstand.

2.15 Material Flexibility

Gerwin(1986) defines this as the ability to handle uncontrollable variations in the composition of dimensions of the parts being processed. It also includes the ability to handle more than one kind of substance either for the same component or different components. The need for this flexibility arises mainly because of the uncertainty as to whether the material inputs to a manufacturing process meet the required standards. Hence as can be expected this is associated with the abilities of the workers and equipment to adjust for the unexpected variations in the inputs.

2.16 Sequencing Flexibility

Gerwin(1986) defines this as the ability to rearrange the order in which different kinds of parts are fed into the manufacturing process. The uncertainty in the delivery times of raw materials gives rise to the need for a system to possess sequencing flexibility.

2.17   Demand Flexibility

Gupta and Goyal(1989) [19]define this as the adaptability to changes in demand. It can be measured in terms of inventory cost of the finished product and the raw material. It is given as the ratio of the physical output of the system to the inventory cost of finished product and raw material. Gustavsson(1984) [22] and Son and Park(1987) have define demand flexibility in such a way that it is quite similar to the definition that Browne et al give of volume flexibility.

2.18   Mix change flexibility

Carter(1986) [7] defines this as the ability of the system to change the product mix inexpensively and rapidly. This requires that the system respond rapidly to the changing market demand. He also proposes to measure mix change flexibility by

1. the number of different parts that can be produced in the system.

2. the range of change in the product mix while maintaining efficient production

3. the cost of making the changeover.

4. time required to changeover.

2.19   Product flexibility

Browne et al. (1984) [3] define this as the ability of change to process new part types. A similar definition is also given for parts flexibility in Taymaz(1989) [10] as the ability to add and remove parts from the mix over time.

Product flexibility has also been called as the changeover flexibility by Gerwin [11]. This is necessitated because of the uncertainty as to the length of the product life cycle.

Son and Park(1987) [8] define product flexibility as the adaptability of a manufacturing system to changes in the product mix. Product flexibility for a given period F_P is given as

$$F_P = O_T/A$$

where A is the setup cost. This is defined in Sethi and Sethi [9] as the ease with which new parts can be added or substituted for existing parts. Thus this actually is a measure of the responsiveness of the firm to the market changes. This has also been called as the part mix flexibility by Chatterjee et al(1984) [13].

2.20   Environmental flexibility

This is defined as the ability of the system to withstand disturbances due to external factors. There are many methods by which the manufacturing systems is affected by its wider environment. Changes in market demand, changes in technology, evolving social attitudes, the political environment, the status of the economy are all factors which affect the manufacturing system.

2.21 Production Flexibility

Browne et al. (1984) [3] define this as the universe of part types that can be processed. Sethi and Sethi(1990) also define this similarly but they add the rider that no major equipment be added to the system. This flexibility is important in that it allows the introduction of new products in a relatively short period of time and at a reduced cost.

Mandelbaum(1978) calls it as the state flexibility which relates to situations where a given system is able to operate well in many different circumstances. Gerwin(1982) and Frazelle(1986) have called this as the mix flexibility in contrast to some other authors who have given a definition of mix flexibility which makes it similar to that of process flexibility. Zelenovic(1982) gives a similar definition to what he calls as the application flexibility.

2.22 Expansion Flexibility

Browne et al.(1984) [3] define this as the ability to easily add capability and capacity. Sethi and Sethi(1990) define Expansion flexibility as the extent of overall effort needed to increase the capacity and capability of a manufacturing system when needed. This actually gives a measure of the time and cost required for the launch of a new product or for adding extra capacity in case of existing products.

Azzone and Bertele(1987) define expansion flexibility as the number of product mixes that the system can produce by adding new machines. Gustavsson(1984) [22] gives it a different name as the machine flexibility. This though as defined by Browne et al. is different in the sense that in the present case we are not concentrating upon the actual operations themselves. This has also been called as the long-term flexibility by Warnecke and Steinhilper(1982) [23].

2.23 Design change flexibility

According to Bernardo and Zubair(1992) [14] Design change flexibility is concerned with the fast implementation of engineering design changes for a particular part.

For a system to possess design change flexibility the product design, process planning and manufacturing functions are to be integrated. Thus this permits the rapid and inexpensive implementation of engineering design changes for a particular part. A somewhat related measure is the produce flexibility given in Azzone and Bertele(1987). They define it as the ability of the system to produce new products with minimal cost.

2.24 Configuration flexibility

Configuration has been defined by Browne et al (1984) [3]as the ease of modification of the transportation system as new centres are added to the FMS or as new products are added requiring new station to station movement.

2.25 New product flexibility

Dixon (1992) [1] defined new product flexibility as the ability to introduce new products. Slack(1983) has also given a definition of new product flexibility but it is similar to the definition of product flexibility as given by Browne et al.

2.26 Modification flexibility

Dixon(1992) [1] defined this as the ability to better meet customer needs by modifying existing products. This arises because of the uncertainty as to the particular attributes demanded by the customer in a product.

This can be measured in terms of the number of design changes made in a component per time period. Thus provision of modification flexibility requires that the workforce be quickly able to modify the operating procedures. The equipment also ought to be such that refixturing is facilitated.

2.27 Application flexibility

Zelenovic (1982) [20] defines this as the value of design adequacy. Design adequacy is the probability that the given structure of a production system will adapt itself to environmental conditions and to the process requirements, within the limits of the given design parameters.

2.28 Quality flexibility

Slack(1983) defines this as the ability of the system to change the quality level of it's products. The available process technology and the workforce skills limit this flexibility. This definition seems to be quite similar to that given by Browne et al for process flexibility.

3 Developing an overall flexibility measure

For developing an overall measure of F.M.S. flexiblity four factors are suggested. These factors integrate all the flexibility types related earlier. Table 1 shows the categorization of all the flexibility types into four factors.
Flexibility = f(Product, Product mix, Process, Environment)
F = f(P, PM, Pr, E)

All the flexibility measures described earlier can be grouped into either of the above four categories. For eg it can be seen that design change flexibility, configuration flexibility, new product flexibility, modification flexibility,volume flexibility,equipment flexibility as well as production flexibility are influenced to a great extent by the product flexibility inherent in the system. Configuration and modification flexibility are influenced also by the product mix flexibility. Thus each of the flexibilities defined in the earlier section can be written as a subset of the four main flexibilities given earlier in this section. Similarly machine flexibility, routing flexibility, operation flexibility and adaptation flexibility are reflected by a measure of the process flexibility. The grouping of all the flexibility measures described earlier into 4 categories is shown in figure 1.

We consider the case whereby the emphasis placed on each of the above elemental flexibilities can be varied by changing the corresponding weights. Let w_i be the weight placed on the ith element in the equation given above. Then a measure of flexibility can be written as

$F = \sum_i w_i f_i$

i-> the elements 1, 2, 3, 4 corresponding to P, PM, Pr and E.

f_i -> the flexibility value for element i

But a problem in using the above equation directly is that f_i has to be determined. In the following we suggest a method whereby each of the above four grouped flexibilities are scored. The actual score of the system is then taken as f_i.

3.1 Product flexibility

The life cycle of a typical product can be depicted typically by the figure shown above. It can be divided into 5 phases as shown above. Different demands are placed on the flexibility of the manufacturing system in the different stages.

During the birth stage the system should be capable of incorporating the different design stages at the small volumes required both quickly and economically. The growth stage is characterised by a continuously increasing volume of the final product manufactured. During the maturity and saturation phases growth changes are small. Emphasis is laid on the manufacturing system being able to produce at high volume and competitive price because of fierce competition. Design changes are negligible. Flexibility may be required to overcome short term fluctuations. Finally during the decline stage the manufacturing system has to produce smaller volumes economically. The system should also be capable of producing its replacement with a minimum of change. Hence keeping these aspects in mind a flexibility scale is given as

1 -> Changes during the maturity stage are easily incorporated.

2 -> Changes during the maturity and saturation stages are easily incorporated.

3 -> The manufacturing system is capable of overcoming any changes occuring during the maturity, saturation and decline stages.

4 -> The manufacturing system is capable of overcoming any changes occuring during the maturity, saturation, decline and growth stages.

5 -> The manufacturing system is capable of overcoming any changes occuring during the maturity, saturation, decline, growth and birth stages.

It has to be pointed out here that the actual ranking of the manufacturing system depends on what the manufacturing manager perceives can be handled by the manufacturing system during the life time of the product.

Category	Flexibility	
Product flexibility	Volume flexibility Production flexibility Configuration flexibility Modification flexibility	Equipment flexibility Design change flexibility New product flexibility
Product mix flexibility	Customizing flexibility Production flexibility Configuration flexibility	Mix change flexibility Expansion flexibility Modification flexibility
Process flexibility	Machine flexibility Operation flexibility Path flexibility Quality flexibility Material handling flexibility	Routing flexibility Adaptation flexibility Sequencing flexibility
Environmental flexibility	Routing flexibility Capacity flexibility Material flexibility Demand flexibility	Volume flexibility Delivery flexibility Sequencing flexibility Application flexibility

Figure 1: Classification of the different flexibilities

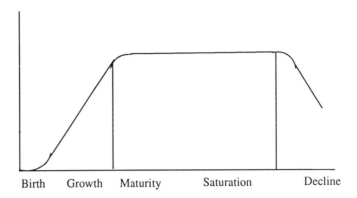

Birth   Growth   Maturity   Saturation   Decline

Figure 2: A typical product life cycle

3.2 Product mix flexibility

This is the ability to manufacture different products at the same time and as such is concerned with the number of different products being manufactured by the system at any time, variations in the product size, shape etc. The system's ability to switch quickly from one product to another is also important. Hence a flexibility scale can be given as

1 –> Change of product not supported at all

2 –> Product variations allowed only for products belonging to the same family.

3 –> Product variations allowed for products belonging to related families.

4 –> Different combinations of products allowed but problems persist as regards volumes, switchover etc.

5 –> Any combination of products, in any volume with switch being done very quickly with very little cost.

3.3 Process flexibility

This reflects the ability of the system to produce components and assemblies in different ways. Thus this minimizes the effect of machine or system breakdown and also makes scheduling easier. Hence a measure of flexibility can be

1 –> Transfer line

2 –> Assembly lines

3 –> Dedicated general purpose machines

4 –> Standard general purpose machines

5 –> Job shop or machining center

3.4 Environmental flexibility

This is basically concerned with how the system can cope with the disturbances generated by the external environment. A scale to measure this can be

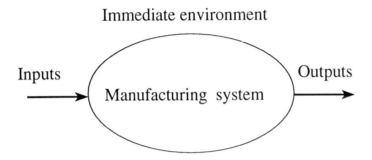

Figure 3: A typical manufacturing system

1 -> Changes neither in the immediate environment nor in the far environment are supported.

2 -> Changes in the far environment only are overcome.

3 -> Changes in some parts of both the far and immediate environments are overcome.

4 -> Majority of the changes in the immediate environment are supported.

5 -> Nearly all the anticipated disturbances which can occur in the immediate and the far environments can be overcome by the manufacturing system.

3.5 Overall Flexibility assessment

A score sheet has been developed for measuring flexibility of FMS and is shown in Figure 4. Relative importance for each category can be incorporated by determining w_i using delphi method or any other technique like A.H.D. This score can be used to compare various FMS before acquiring one.

4 Conclusions

It is quite obvious that there is no uniformity in the literature as far as the various definitions of the elemental flexiblities are concerned. Hence in this paper we have looked at the different definitions of flexibility as given in the literature. The conflicts and similarities between different definitions have also been pointed out. Another problem facing managers of manufacturing in recent years is the inability to come up with a quantitative measure of the level of flexibility present in the manufacturing system. Hence in this paper we have also concentrated on how to come up with an integrated measure of the level of flexibility inherent in any given manufacturing system. This should be of much help to the managers in deciding the present level of flexibility and decide upon whether they would need any increase in the flexibility level. The proposed

82

Factor	Factor weight (w_i)	Characteristic	Score (S_i)	w_iS_i
Product flexibility		Can changes during maturity stage be easily incorporated	5	
		Can changes during maturity and saturation stages be easily incorporated	10	
		Can changes during maturity ,saturation and decline stages be easily incorporated	15	
		Can changes during maturity ,saturation, decline and growth stages be easily incorporated	20	
		Can changes during maturity ,saturation, decline,growth and birth stages be easily incorporated	25	
Product mix flexibility		Is change of product not supported at all	5	
		Are product variations allowed only for products belonging to the same family	10	
		Are product variations allowed for products belonging to related families	15	
		Are different combinations of products allowed but problems persist as regards volumes, swithover etc.	20	
		Is any combination of products allowed, in any volume with switch being done very quickly with very little cost.	25	
Process flexibility		Transfer line	5	
		Assembly lines	10	
		Dedicated general purpose machines	15	
		Standard general purpose machines	20	
		Job shop or machining center	25	
Environmental flexibility		Are changes neither in the immediate environment nor in the far environment supported.	5	
		Are changes in the far environment overcome.	10	
		Are changes in some parts of both the far and immediate environments overcome	15	
		Can a majority of changes in the immediate environment be supported	20	
		Can nearly all anticipated disturbances which can occur in the immediate and the far environments be overcome.	25	
			$\Sigma w_iS_i=$	

Figure 4: Score sheet to measure manufacturing system flexibility

measure can also guide the manager in deciding about which aspects of flexibility he should concentrate upon while expanding the manufacturing system.

References

[1] Dixon J. R. Measuring manufacturing flexibility: An empirical investigation. *European Journal of Operational Research* ,, pages 131–143, 1992.

[2] Slack N. Flexibility as a manufacturing objective. *International Journal of Operations and Production Management*, 3:4–13, 1983.

[3] Browne J. Dubois D. Rathmill K. Sethi S.P. and Stecke K.E. Classification of fms. *FMS magazine*, pages 114–117, 1984.

[4] Buzacott J.A. The fundamental principles of flexibility in manufacturing systems. *R. Lindholm (ed.) Proceedings of the First International Conference on Flexible Manufacturing Systems*, pages 13–22, 1982.

[5] Das S.K and Nagendra P. Investigation into the impact of flexibility on manufacturing performance. *International Journal of Production Research*, 31(10):2337–2354, 1993.

[6] Barad M. and Sipper D. Flexibility in manufacturing systems-definitions and petrinet modelling. *International Journal of Production Management*, 26(2):237–248, 1988.

[7] Carter. Designing flexibility into automated manufacturing systems. *Proc.II ORSA/TIMS conference on FMS: OR models and applications*, pages 107–118, 1986.

[8] Son Y.K and Park C.S. Economic measure of productivity, quality and flexibility in advanced manufacturing systems. *Journal of Manufacturing Systems*, 6(3):193–207, 1987.

[9] Sethi A.K. and Sethi S.P. Flexibility in manufacturing: A survey. *International Journal of Flexible Manufacturing Systems*, 2(4):289–328, 1990.

[10] Taymaz E. Types of flexibility in a single machine production system. *International Journal of Production Research*, 27(11):1891–1899, 1989.

[11] Gerwin D. An agenda for research on the flexibility of manufacturing processes. *International Journal of Operations and Production Management*, pages 38–49, 1986.

[12] Frazelle E.H. Flexibility: A strategic response in changing times. *Industrial Engineering*, pages 17–20, 1986.

[13] Maxwell W Chatterjee A., Cohen M. and Miller L. Manufacturing flexibility: Models and measurements. *Proc. First ORSA/TIMS special interest conference on FMS, Ann arbor , MI*, pages 49–64, 1984.

[14] Bernardo J.J. and Mohammed Zubair. The measurement and use of operational flexibility in the loading of flexible manufacturing systems. *European Journal of Operational Research,*, pages 144–155, 1992.

[15] Azzone G. and Bertele V. Comparing manufacturing systems with different flexibility : A new approach. *Proc. The Decision Sciences Institute Boston*, pages 690–693, 1987.

[16] Kusiak A. Application of operational research models and techniques in flexible manufacturing systems. *European Journal of Operational Research*, 24:336–345, 1986.

[17] Graves S. Safety stocks in manufacturing systems. *Journal of Manufacturing and Operations Management*, 1:67–101, 1988.

[18] Kumar V. On measurement of flexibility in flexible manufacturing systems: An information theoretic approach. *in K.E.Stecke and R.Suri (eds) Proceedings of the second ORSA/TIMS Conference on Flexible Manufacturing Systems: OR models and applications, Elsevier, Amsterdam*, pages 131–143, 1986.

[19] Gupta Y.P and Goyal S. Flexibility of manufacturing systems: Concepts and measurements. *European Journal of Operations Research*, 43:119–135, 1989.

[20] Zelenovic D.M. Flexibility- a condition for effective production systems. *International Journal of Production Research*, 20(3):319–337, 1982.

[21] Slack N. and Correa H. The flexibilities of push and pull. *International Journal of Operations Management*, 12(4):82–92, 1992.

[22] Gustavsson S. Flexibility and productivity of complex processes. *International Journal of Production Research*, 22(5):801–808, 1988.

[23] Warnecke and Steinhilper. Flexible manufacturing systems, edp planning; application examples. *in R. Lindholm(ed) Proceedings of the first International Conference on Flexible Manufacturing Systems, North Holland Amsterdam*, pages 345–356, 1982.

Flexible Manufacturing Systems: Recent Developments
A. Raouf and M. Ben-Daya (Editors)
1995 Elsevier Science B.V.

Flexibility and Productivity in Complex Production Processes

Sten-Olof Gustavsson

Department of Industrial Management
Chalmers University of Technology
Sweden

Abstract

The drastical changes in market demands, and the rapid technological developments, has created a need for:

- more flexible production systems

- more complex products with a larger degree of variation.

There is a strong force towards the use of more and more mechanized and automatized equipment, from single NC-machines to complete manufacturing systems. At the same time there is a need for flexibility towards changes of the products. These changes have to be made in a limited time, and without the need of large reinvestments in the production system.

This means that there more often must be a dicsussion regarding flexibility versus productivity before the production system is designed. I will discuss

- methods for calculation of different flexibility levels

- strategies for a more flexible view upon products and processes

- examples and results from different areas within the Swedish industry.

1   Introduction

As a result of increased industrial automation and of the trend towards an ever shorter life cycle for a product, it has become apparent that the flexibility of the machinery needed for complex production processes is now of overriding importance for long-term profitability.

The danger exists that a short-term gain in production is achieved by using machinery or equipment that then becomes redundant on the introduction of a new model. There might thus be a conflict of aims between flexibility and productivity. Strategically speaking, production should be so flexible that neither the product or the renewal of the processes should be hindered by "sunk" costs in production.

2   Productivity

This concept is familiar and has been dealt with in a variety of contexts so only a few aspects will be touched on here very briefly.

Productivity

- corresponds well with mass production (over a long period)

- brings the short-term perspective into focus

- draws attention to internal (the production apparatus) rather than to external questions (what the client judges as valid i.e. the right product at the right price).

Albeit simple in theory, the concept of productivity is beset by complications and difficulties in practice since how it may be interpreted will depend on factors such as the time aspect, product development, inflation and econometrics.

Considerable advantages can be gained by utilizing straightforward simple measures as work productivity in physical measures of quantity per employee, output productivity and capital productivity without coupling this to the total productivity index.

Key figure comparisons with competitior will help to give a very comprehensive view of the extent and power of the competition. Obviously, in every industry, there must be a continuous follow-up of production, preferably with the help of several key figures. Assessment, for example, of a 10 percent productivity increase should be a familiar routine in every business.

It is equally obvious that continuous comparison should be made with other manufacturers within or outside the company. The real challenges, however, arise every time a decsion is made for the future, which is to say, that action should be taken on the basis of accumulated experience. The investments in all these resources are expected to pay off (at least in the long-term). This means that the use of these resources will often make demands on both productivity and flexibility, and once more, the needs of the market are decisive.

3   Productivity and Flexibility

Production (i.e. goods) is defined as the manufacture of products with the help of personel, material, equipment (hard and soft-ware) and capital. The consumption of resources is compared with earlier consumption in budget control and other steering instruments.

Products are subject to changes:

- a change of technology (electronics take over from mechanics),

- "rationalisation" (one component does the work of several),

- changes in fashion

A company ultimate success depends on its ability to utilize resources and meet the needs of the market. These internal factors steer demand and in turn the volume of business and the price of the commodity.

In addition to all this, there must be flexibility in respect of external factors. These may be:

- fluctuations of the market

- seasonal fluctuations,

- competition from other companies

There is also another interesting dimension to take into account, namely the question of the product's life cycle. Uncertainty is always present from the moment a product is introduced on the market (will it be a success or not?) to the end (when will the market vanish?) .

4   Flexibility

Flexibility comes from the Latin word for bendable. Other expressions are adjustable and mobile. Industrially speaking the word means adaptable and capable of change. The concept has been a subject of interest to both production engineers and research workers. The flexibility of the work group has been examined by, for example, Kozan (1982) and the flexibility of the manufacturing system by Hjelm (1982). Warnecke et al (1981) have discussed the flexibility of the whole production system.

Flexibility can be defined as follows:

1. Changes in the product

   - improvements, new components,
   - several variants

2. Changes in the production system

   - new machinery and production methods,
   - new systems (for example, computerization),
   - new personnel

3. Changes in demand

   - insecurity over a period,
   - fluctuations (over the year, for example).

It can be said that these three types of changes made demands on the production system as regards both the short-and long-term view. I can make the following basic divisions according to the time aspect involved when it comes to assessing productivity versus flexibility.

1. Operational problems. Short-term - for example, having to replan in order to cope with a breakdown of vital machinery, or an unexpected shortage of material.

2. Tactical problems. Medium-term, such as changes in design or rate of production.

3. Strategic problems. Decisions with long-term effect such as investments in machinery of expansion.

I have three examples to illustrate the differences.

A. Shipyard
   Ships - alternative production
B. Hobby mowers - snow scooters
C. Car factory
   Line-out

It is essential to identify "the critical time perspective or perspectives", that is, to be ready to cope with the operational, tactical and/or strategic problems. This analysis, of course, should be made preferably whe the system is constructed but since external circumstances constantly alter, there should be discussion of these matters at regular intervals.

If the analysis is to be valid it must look closely at only the production system's resources (personal, machinery, etc) but also at the qualities of the product or products themselves. Which of the above is crucial for success? The answer would seem obvious: the product. No customer - no business, which means that a thoroughly attractive product is the be-all and end-all of the matter. However, attractiveness is not just a question of function, it is also one of price and quality. There is thus a clear connection between the product and the production system with its resources.

In conclusion, I would like to point out that is every production system the following must be decided.

1. wich level is primarily critical (for the company and sections of the company) operational, strategic, tactical,

2. which resource (personnel, machinery/system or product) is primarily critical and therefore in need of particular attention.

It is only after the crucial factors have been identified that the work of constructing an effective (productive and flexible) production system can begin.

I should like to illustrate our reasoning with another example which is interesting in that it demonstrates all three types of flexibility seen from the strategic, tactical and operational angle.

Example: Truck manufacutre.

A case study of this type can always be said to be unique but one should bear in mind that:

1. the same items is not so common but products serve the same purpose, the only differnce between them being the design of a few details.

2. Variations of the same basic construction can be designed in different ways.

The list could be made longer. It is enough to say that all production systems are unique, some more unique than others. Naturally, a concept such as flexibility (or productivity) cannot be expected to follow a simple standard pattern. Let us instead indicate approaches to the probloem.

F		
L	Capacity	- Change of volume
E	Product	- Design, models, generations
X		
I	Steering system	- Structural programme, raw material/
B		primary products
I	Production	- Direction, flow
L		
I	Machinery	- Machines, tools, fixtures
T		
Y	Personnel	- Competence, structure

Figure 1. Structures in the concept of flexibility.

Good examples of flexible equipment are the NC-machines which have reduced the rigging or starting-up time for machines. This has allowed for economical small-scale production where previously multi-specialized operation in several installations reqired large-scale series for manufacture to be economically viable. A good example of product flexibility is a construction composed of modules whose end-product, although made up of a unique combination of modules, is itself composed of mass-produced components. A pizzeria is another example.

Good flexible capacity can be achieved by utiizing the parallel principle, whereby a parallel product line can be added or discontinued as demand varies. Good examples, by all means, but it must be possible to measure how good they are. To do this requires inventiveness. Here are a couple examples:

1. Flexibility of machinery can be measured as the ratio of the investment's residual value for the next product model to the original investment, i.e. an index between 0 and 1.

2. Product flexibility can be measured as the ratio of the residual value of the old model to the new model divided by the original value for the old model.

5 Strategies for the assessment of flexibility versus productivity

There are at least three ways of calculating so-called "optimal" flexibility.

1. All starting-up costs and all other costs are grouped under "life-cycle cost" to facilitate optimization.

2. Only model-restricted machinery is optimized in the "life cycle" sense.

3. Expensive process machinery is made to be used irrespective of model and has been standardized to the extent that it can be used generally. To take an example: the Swedish Match Company by the 1920's had already standardized its matchstick to the point that the company could risk building fully automated machinery for mass production of the matchstick at the same time that it could offer a large assortment of shapes and sizes for the boxes and labels.

Model I has been applied in the aviation industry where so-called "break-even" calculations have been made including all costs; product development, prototypes, testing, grounds, buildings, tools ligs, fixtures etc. an estimate is then made of the size of the prospective market, for example 300 [planes. It is then a simple matter to evaluate the result of bringing in more mechanized and automated equipment. A sav ng of 1000 Swedish crowns in working costs per plane is the equivalent of a maximum investment of 300,000 crowns.

In principle, the model is simple but the limit for the total legth of the series must be set so low that it is certain to be exceeded and the model gives priority to alternatives bringing in a profit.

Model II is applied in the motor/automotive industry, where costs are separated into model-restricted and model-free categories. If all doubtful costs are classified as model-restricted, it is possible even here to calculate the value of, for example, work-saving automated machinery. A saving of 100 Swedish crowns in work costs is the equivalent of a maximum investment of 100 million crowns for a total life cycle length of 1 million cars.

In the April 5th, 1982 since of "Fortune" Porsche's car body assembly was described as flexible, manual with 25 man-hours per car. A more normal plant for a series length of 250,000 cars a year would require 5 man-hours pe car. Since Porsche only produces about 10,000 of the 911 model per year, the difference would mean a capital cost of 2,000 Swedish crowns per car, a total of 20 millin crowns a year. A possible investment of roughly 80 million crowns in all for mechanized machinery. This is probably not feasible at this cost. If one reckons on a volume of 250,000 cars per year, the value of 20 man-hours saved is roughly 500 million crowns per year and a total possible investment of 2,000 million crowns, which could be done with good profit. The conclusion is that it is easier to achieve flexibility with small volumes for economic reasons.

These examples shows that rough estimates can be made which will serve for the assessment of different production systems. Unfortunately, this seldom happens, possible owing to a widespread belief that the higher the degree of flexibility the lower the level of productivity. (cf. Warnecke et al. 1981).

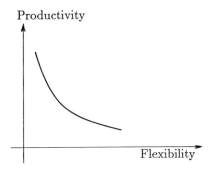

Figure 2. Productivity vs. flexibility.

It is also commonly believed that increased flexibility must involve higher investment costs.

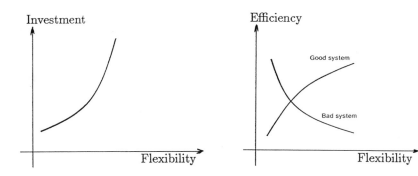

Figure 3. Investment vs. flexibility.   Figure 4. Efficiency vs. flexibility.

The relationship between flexibility and efficiency is heavily dependent on the structure of the chosen system and so the design of the system becomes instrumental to the success or failure of the relationship.

Which strategy for flexibility?

Demand Flexibility

If all the analyses indicate that there is little risk of significant changes in demand, then investment can be made in a process which constitutes a total solution for a fixed capacity. Generous margins will ensure a certain amount of flexibility. However, if there is a risk that demand will vary significantly, and if the total demand is hard to estimate, then investments should be made in small parallel steps which can easily be added to or stopped short.

Product flexibility

If there is a likelihood that there will be frequent changes of model and product, then the product should be composed of modules with allowance made for step-by-step changes. It is, of course, risky to invest in module-restricted process machinery for suchlike products.

Flexibility of machinery

If it is very likely that there will be a technological change from, for example, steel to plastic, then it is advisable to employ subcontractors who have several other customers with other products which facilitate the technological transfer.

6 Conclusion

A consciously planned flexibility can be achieved without significant increases in investment. The following guide lines can serve for the design of both product and production systems.

The use of modules

Many alternative machinery installations can be converted if each module has access to air, electricity and compressed air, for example. Standard components can be combined as larger pieces of mechanical equipment. Products can be renewed successively if they are composed of modules. ASEA, for example, constructed switch gear by using modules.

Standardization

Well-planned rigorous standardization will facilitate the introduction of larger series, improve economy and encourage demand. The Hasselblad system, for example, was constructed with a number of rigorously standardized measurements which can be retained more than 30 years later.

The Variant Tree

Initial operations in a process should not include variants. These, however, can be produced economically if the final operations encompass a large number of alternatives.

Development towards automation

The requirement that all automated machinery can be quickly adjusted to new dimensions promotes flexible machinery. Rapid changes in the machinery allow for small series and foster a large capital turnover. Life-cycle costs have become increasingly relevant to the decision-making process provided that the aim is to review the cost of the whole life-cycle for a product and/or production system. It is necessary, however, to include appraisals of the development of volume, inflation and shifts of balance in the cost relations.

Last but not least, there is a need for efficient personnel at all levels. This means they should be both productive (interested in making improvements and flexible (able to revise their thinking). Every manager's primary task is to create such personnel.

Acknowledgements
This paper has been prepared in collaboration with my close friend Bengt Almgren to whom I wish to express my gratitude.

References

[1] B. Colding Productivity gains through the realization of the integrated manufacturing system today in limited manpower production at low cost.

[2] S. Hjelm: Flexibla Automatiserade tillverkningssystem - Mal och forutsattningar i Sverige. The Royal Institute of Technology, Stockholm, Sweden, 1982.

[3] K. Kozan: Work Group Flexibility: Development and Construct Validation of a Measure. Human Relations, Vol. 35, N. 3, 1982, pp. 239-240.

[4] H.J. Warnecke, H.J. Bullinger, J.H. Kolle, German manufacturing industry approaches to increasing flexibility and productivity. World productivity conference Detroit 1981.

[5] S.O. Gustavsson, Motive forces for and consequences of different plant size VI International Conference on Production Research Novi Sad 1981. Fortune, April 5, 1982 Automaking on a human scale, p. 87-93.

Flexible Manufacturing Systems: Recent Developments
A. Raouf and M. Ben-Daya (Editors)
1995 Elsevier Science B.V.

Flexibility in Pull and Push Type Production Ordering Systems - Some Ways to Increase Flexibility in Manufacturing Systems

R. Muramatsu, K. Ishi and K. Takahashi

Department of Industrial Engineering School of Science and Engineering,
Waseda University 3-4-1 Ohkubo, Shinjuku-ku, Tokyo 160, Japan

Abstract

In recent years, many flexible manufacturing systems have been developed [1,3, 4]. FMS is important in order to adapt to severe changes in market conditions and technology and to increase productivity.

This paper introduces variables for evaluating flexibility and discusses ways to improve the flexibility of single-stage production systems with different characteristics. However, since most production systems are multi-stage mixed type systems, this paper also shows how the "amplification" or increase in the variability of multi-stage production systems is an important factor affecting flexibility and productivity. Ways of reducing the "amplification" by using production ordering systems are also presented in this paper.

The multi-stage production ordering models discussed here are simplified. They neglect characteristics of single production systems which comprise total multi-stage production systems. The models also neglect transportation and material flow parameters.

The flexible manufacturing system of today is treated as one stage which is part of an overall system consisting of a succession of production stages with different characteristics. However, it should also be thought of as a way to increase the flexibility and productivity of the system as a whole.

1   Introduction

There are two major areas for consideration by management. The first is the marketing area. Cultivation of new markets and changes in user needs require new products. Market segmentation increases the number of product specifications. Increases in the number of products create greater fluctuation in each product's demand, and product life cycles grow shorter. The second is the production area in which product lines and specifications are varied, parts configurations are complex and the precision and speed of the production equipment in each stage differ according to changes in technology.

Moreover, most industries have many successive stages of production, inventory and material handling. In these cases, the higher the stage, the longer the total lead time required from the material processing stage to the final assembly line.

Management needs to adapt production systems to market and technology changes:

In order to adapt production systems to changes in the market and production areas, production management must consider the following:

1. Minimization of total lead time from material processing to shipping and minimization of total cost.

2. Minimization of time needed to change to new products.

3. Keep down the amount of lead time and keep the inventory at the same level in each stage, regardless of increases in the variety of products.

4. Keep variations in total work load and inventory at the same level even if the fluctuation in demand for each product increases.

5. Keep variations in the production ordering and inventory levels the same in the face of forecasting error and down time.

6. Keep the "amplification" in the production ordering and inventory level in preceding processes the same.

7. Control production factors which increase total cost.

Production systems to satisfy management's needs and increase flexibility:

There are many ways to satisfy management's needs for each production type. There are also many different opinions regarding the concept of "flexible manufacturing systems" [5-12].

The concept of the "flexible manufacturing system" is defined in this paper as a system which satisfies management's need to minimize resources and time as mentioned above. The variables used to evaluate flexibility are the number of parts and products, the length of total lead time, quantity of inventory and the "amplification" in production and inventory level in each stage of the system under the constraints of investment and product costs.

In this paper, the production types are classified into single production systems and multi-stage production systems. Further, single stage production systems are classified into machining process systems, lot production systems and assembly line production systems according to differences in production scheduling, equipment and production methods. The multi-stage production systems are classified into multi-stage mixed types of machining processes, multi-stage lot production systems and multi-stage machining, lot and assembly line production systems.

2 Factors increasing flexibility in single stage production systems

In single stage production systems, there are many factors which increase flexibility within each production type.

In the machining systems, in order to adapt to a variety of parts or products, it is necessary that the set up time (t_s) and the machining time (t_m) are shortened for minimizing lead time. And it is necessary that the feeding and moving out time (t_h) are shortened for minimizing inventory quantity. Thus, realizing investment saving depends on the use of alternative equipment. Labor cost savings may be realized by automation and robotization.

Adapting lot production systems to handle an increased variety of parts or products requires shortening the set up time and making smaller production lot sizes (q). Set up time (t_s) and processing time (t_M) must also be shortened in order to make larger production lot sizes which minimize lead time. It is also necessary to shorten set up time to minimize inventory quantity. The realization of investment savings depends on the use of alternative equipment. Labor costs saving can be realized through automation and robotization.

Assembly line production systems change to multi-model mixed assembly line production systems in order to adapt to a variety of products and to minimize the total lead time and inventory level required. Investment in the systems can be reduced through equipment overlapping. Costs can be reduced through inventory savings. Labor costs would be saved by increasing the ratio of versatile operators. [13].

Factors increasing the flexibility in the single stage production systems are shown in Table 1.

3 Factors increasing flexibility in multi-stage production systems

Concerning the flexibility of multi-stage production systems, the key problem is to prevent an increase of "amplification" in production ordering and inventory level from final production stage to the stages preceding final production [14].

The "amplifications" are defined by equation (1).

$$Amp(O^k) = V(O^k)/V(D)$$
$$Amp(B^k) = V(B^k)/V(D) \tag{1}$$

where, $V(O^k)$: quantity variance upon production ordering at the k-th stage,
$\quad V(D)$: market demand variance,
$\quad V(B^k)$: inventory variance at the k-th stage.

Therefore, desirable systems are as follows;

$$1.0 \geq Amp(0^1) \geq Amp(0^2) \geq Amp(0^3) \cdots \geq Amp(0^k)$$

Table 1
Factors Affecting Flexibility in Single Stage Manufacturing Systems

Evaluative variables of Process Types of Process	Variety of Parts or Product	LT Minimization	Inventory Minimization	Investment Savings	Cost Savings
Machining Production	• $t_s \to$ min	• $\begin{cases} t_s \\ t_m \end{cases} \to$ min	• $t_h \to$ min	• Depends on equipment	• Labor cost savings realized by automation & robotization • Depends on investment cost
Lot Production	• $t_s \to$ min • $q \to$ min	• $\begin{cases} t_s \\ t_m \end{cases} \to$ min • $\to$ max	• $\begin{cases} t_s \\ t_h \end{cases} \to$ min • $q \to$ max	• Depends on equipment	• Labor cost savings realized by automation & robotization • Depends on investment cost
Assembly Line Production	• Change to multi-model mixed assembly line production system • Versatile operators ratio $\to$ max • $t_m \to$ min • Number of work stations $\to$ min • Sequencing			• Savings due to overlapping equipment	• Saving labor and inventory

Notation t_s : set up time t_m : manufacturing time per piece
     t_h : handling and transportation time q : production lot size
     versatile operator ratio = (total number of work stations mastered by each worker)
                / { (number of work stations) * (number of workers)}

$$1.0 \geq Amp(B^0) \geq Amp(B^1) \geq Amp(B^2) \cdots \geq Amp(B^{k-1})$$

where, the first stage is a final production or an assembly stage and the k-th stage is a raw material processing stage. A change for the worse in $Amp(0)$ and $Amp(B)$ is serious, causing increases in lead time, inventory levels and costs.

In machine processing systems of multi-stage production systems, preventing large increases in lead time and inventory is needed to shorten processing and set up time. Consideration of job shop scheduling problems on multi-job and multi-process, group technology (GT), substitute machinery and the ratio of versatile operators is also important in order to try to adapt the system to management's needs.

In the multi-stage lot production systems, the prevention of large increases in lead time and inventory are needed not only to minimize set up time, processing time and feeding and moving out time but is also needed to hold the ordering system to the optimal lot size, for minimizing total lead time, for preventing work congestion between one stage and the immediate preceding stage, and for increasing flexibility by utilizing the ratio of versatile operators and available buffer systems [15,16].

In the multi-stage mixed type production systems, consisting of machining processes, lot production processes and multi-model mixed assembly line systems [17], the available buffer systems and the ratio of versatile operators are needed to increase flexibility. Factors which increase flexibility in multi-stage production and inventory systems are shown in Table 2.

Reducing set up, processing and feeding and moving out time is accomplished by improvements in equipment and computer-aided engineering. The improvement of ordering systems, determination of optimal lot size and the solving of job sequence problems are accomplished by improving and innovating production management technology and computer-aided engineering. The group technology and designing of multi-model mixed assembly lines are developed through improvements and innovations in equipment, production management and computer-aided engineering.

Thus, the term "FMS" describes the following:

- Flexible Machining Systems - systems which have automatic processing, tooling, loading and unloading, and handling.

- Flexible Manufacturing Systems - systems which are concerned with scheduling and consider the demand of immediately succeeding stages based on consideration of the flexible machining systems.

- Flexible Management Systems - systems which totally optimize the flexible manufacturing systems with the flexible machining systems.

4   A comparative study on the flexibility of two types of ordering systems

There are two types of production ordering systems. One is the "push type" production ordering system and another is "pull type" production ordering system.

Table 2
Factors Affecting Flexibility in Multi-stage Manufacturing System

Evaluative Variables / Types of Succession Process Systems	Variety of Parts or Product	LT Minimization	Inventory Minimization	Amp(0) Minimization	Amp(B) Minimization
Machining Processes	GT	• t_s, t_m and $t_n \rightarrow$ min • Job shop scheduling • Substitute machine • Versatile operator			
Lot Production Processes (two stage)	• t_s, t_m and $t_h \rightarrow$ min • Optimization production lot size in multi-stage $$N^* = \begin{cases} \sqrt{\dfrac{R_1 t_{m_{1,2}}}{\sum_{i=1}^{M} t_{s_{i,1}}}} & (3) \\[2em] \sqrt{\dfrac{R_M t_{m_{M,1}}}{\sum_{i=1}^{M} t_{s_{i,2}}}} & (4) \end{cases}$$			• Desirable ordering system • Available buffers
Mixed Assembly Line, Lot Production and Machining Processes	• t_s, t_m and $t_h \rightarrow$ min			• Desirable ordering system • Sequence of multi-model in final assembly line • Available buffers • Versatile operator

where, N: number of setups; R_i: production quantity of the ith item per period;
$\tau_{i,k}$: production time of; the ith item at the kth stage;
$Amp(0^k) = \frac{V(0^k)}{V(D)}$; $Amp(B^k) = \frac{V(B^k)}{V(D)}$; $V(x)$=Variance of x.
(3) $\sum_{i=1}^{M} \tau_{i,2} \leq \sum_{i=1}^{M} \tau_{i,1}$ and $\tau_{i+1,2} \leq \tau_{i,1}$ (for $i = 1, 2, ..., M-1$)
(4) $\sum_{i=1}^{M} \tau_{i,2} > \sum_{i=1}^{M} \tau_{i,1}$ and $\tau_{i+1,2} \leq \tau_{i,1}$ (for $i = 1, 2, ..., M-1$).

4.1 Concept of "push type" and "pull type" production ordering systems

In the "push type" production ordering system, the ordered quantity in each stage is determined by forecasted demand. Forecasted demand is the length of cumulated lead time from one stage to the final assembly line, and of feedback information of product or in-process inventory in each stage. In this system, the ordered quantity of each production stage is ordered by a central controller. Thus, it may also be called a "centralized ordering system". Material flows are controlled just as if they are "pushed out" from the raw materials stage toward the final stage.

In "pull type" production ordering systems, the ordered quantities in each stage are determined by actual quantities consumed by the immediate downriver stage. Here no central controller is needed. Thus, it may also be called a "decentralized ordering system". Material flows are controlled just as if they were "pulled" into the final product stage from the stages preceding final production.

4.2 General model formulation

Assumptions:

1. Figure 1 and 2 show the schematic diagram of "push type" and "pull type" production ordering systems.

2. The systems consist of K production stages. Each production stage has only one process. Each production process produces M kinds of products.

3. There are two kinds of the inventory stages. The inventory stage $I^{k(i)}$ is the part inventory for the i-th product which has been fabricated by the production stage k. The inventory stage $B^{k(i)}$ is the part inventory for the i-th product which is the on-hand material level for the production stage k. The inventory stage $B^{0(i)}$ is the final product inventory stage of the i-th product.

4. The production lead time for the k-th stage is described as L_p^k. The production quantity ordered to the stage at the end of the T-th period is completed during the $(T + L_p^k)$th period and is stored in the fabricated inventory $I^{k(i)}$ at the end of the $(T + L_p^k)$th period.

5. The handling and transportation lead time from the fabricated inventory $I^{k(i)}$ at the k-th stage to the on-hand inventory $B^{k-1(i)}$ at the (k-1)th stage is described as L_h^k.

6. Back logs are permitted.

7. The raw material inventory supplied for the k-th stage is always sufficient. But at the other stages $1, 2, ..., K-1$, the quantity ordered for production is restricted by the on-hand materials inventory $B^{k(i)}$.

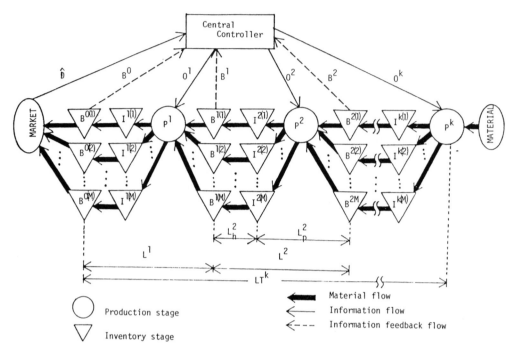

Figure 1: Schematic diagram of the push type ordering system.

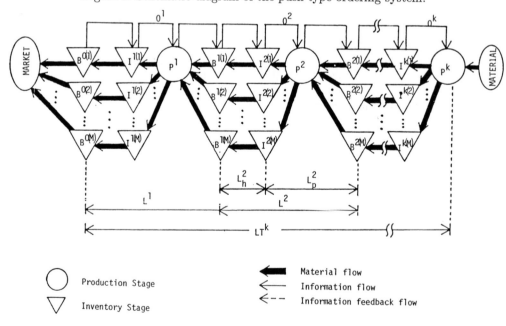

Figure 2: Schematic diagram of the pull type ordering system.

8. The production capacity of each stage is the same as in each planning period.

9. Down time occurs at each production stage. During down time, the production stage stops producing.

Notations:

L_h^k : handling and transportation lead time at the k-th stage,
L_p^k : production lead time at the k-th stage,
L^k : the lead time from production starting time at the k-th stage to the
: handling and transportation completion time at the (k-1)th stage, i.e.

$$L^k = L_h^k + L_p^k \tag{2}$$

LT^k : the accumulated lead time from the final stage to the k-th stage, i.e.

$$LT^k = L_h^k + L_p^k \tag{3}$$

$\hat{D}_{T:T+L}^{(i)}$: the forecasted market demand of the i-th product for the (T+L)th
: period forecasted at the end of the T-th period,

$D_T^{(i)}$: the actual market demand of the i-th product in the T-th period,

$Q_{T:T+L^k+1}^{k(i)}$: the required quantity of the i-th product for the $(T + L^k + 1)$th period
: at the k-th stage calculated at the end of the T-th period,

$O_{T:T+L^k+1}^{k(i)}$: the ordered quantity which is determined on the basis of the required
: quantity with restrictions on production capacity and material inventory.

$P_T^{k(i)}$: the actual production quantity of the i-th product during the T-th
: period in the k-th production stage,

$I_T^{k(i)}$: the inventory of the i-th product fabricated by the k-th production stage
: at the end of the T-th period,

$S^{k(i)}$: the safety stock of the i-th product at the k-th stage,

C^k : the production capacity at the k-th stage during the T-th stage,

X^k : the down time at the k-th production stage during the T-th period,

$e_{T-L:T}^{(i)}$: the forecasting error of the i-th product for the T-th period
: forecasted at the end of (T-L)th period i.e.

$$e_{T-L:T}^{(i)} = \hat{D}_{T-L:T}^{(i)} - D_T^{(i)} \tag{4}$$

$A(i,k)$: the required part quantity of the k-th production stage to manufacture
: one unit of the i-th product.

System equations:

(1) Push type production ordering systems

$$Q^{k(i)}_{T:T+L^k+1} = A(i,k)\hat{D}^{(i)}_{T:T+LT^k+1} + \sum_{\ell=1}^{L^k} A(i,k)\hat{D}^{(i)}_{T:T+LT^{k-1}+\ell} - \sum_{\ell=1}^{L^k} Q^{k(i)}_{T-\ell:T+L^k-\ell+1}$$
$$- B^{k-1(i)}_T + S^{k-1(i)} \tag{5}$$
$$(\text{for } k = 1, 2, ..., k \text{ and } i = 1, 2, ..., M)$$

$$O^{k(i)}_{T:T+L^k+1} = \min\left\{ Q^{k(i)}_{T:T+L^k+1}, \frac{C^k . Q^{k(i)}_{T:T+L^k+1}}{\sum_{i'=1}^{M} Q^{k(i')}_{T:T+L^k+1}}, \right. \tag{6}$$
$$\left. \left(B^{k(i)}_T + P^{k+1(i)}_{T-L^k_n}\right) \frac{A(i,j)}{A(i,k+1)} \right\} \tag{7}$$
$$(\text{for } k = 1, 1, 2, ..., k-1 \text{ and } i = 1, 2, ..., M)$$

$$P^{k(i)}_T = \min\left\{ O^{k(i)}_{T-L^k_p:T+L^k_h+1}, \frac{\left(C^k - X^k_T\right) O^{k(i)}_{T-L^k_p:T+L^k_h+1}}{\sum_{i'=1}^{M} O^{k(i')}_{T-L^k_p:T+L^k_h+1}} \right\}$$
$$(\text{for } k = 1, 1, 2, ..., k \text{ and } i = 1, 2, ..., M) \tag{8}$$

$$I^{k(i)}_T = P^{k(i)}_T \quad (\text{for } k = 1, 2, 3...., k \text{ and} i = 1, 2, 3, ..., M) \tag{9}$$

$$B^{k-1(i)}_T = B^{k-1(i)}_{T-1} + P^{k(i)}_{T-L^{k-1}_h} - P^{k-1(i)}_T . \frac{A(i,k)}{A(i,k-1)}$$
$$(\text{for } k = 2, 3, ..., k \text{ and } i = 1, 2, ..., M) \tag{10}$$

$$B^{0(i)}_T = B^{0(i)}_{T-1} + P^{1(i)}_{T-L^k_h-1} - D^{(i)}_T \quad (\text{for } i = 1, 2, ..., M) \tag{11}$$

(2) Pull type production ordering systems

$$Q^{1(i)}_{T:T+L^1+1} = D^{(i)}_T + \left(Q^{1(i)}_{T-L^1_p,T+L^1_h+1} - P^{1(i)}_T\right) \quad (\text{for } i = 1, 2, ..., M) \tag{12}$$

$$Q^{k(i)}_{T:T+L^k+1} = P^{k-1(i)}_T . \frac{A(i,k)}{A(i,k-1)} + \left(Q^{k(i)}_{T-L^k_p:T+L^k_h+1} - P^{k(i)}_T\right)$$
$$(\text{for } k = 2, 3, ..., k \text{ and } i = 1, 2, ..., M) \tag{13}$$

$$O^{k(i)}_{T:T+L^k+1} = \min\left\{ Q^{k(i)}_{T:T+L^k+1}, \frac{C^k . Q^{k(i)}_{T:T+L^k+1}}{\sum_{i'=1}^{M} Q^{k(i')}_{T:T+L^k+1}}, \left(B^{k(i)}_T + P^{k+1(i)}_{T-L^k_n}\right) \frac{A(i,k)}{A(i,k+1)} \right\}$$
$$(\text{for} k = 1, 2, ..., k-1 \text{ and } i = 1, 2, ..., M) \tag{14}$$

$$P_T^{k(i)} = min \left\{ O_{T-L_p^k:T+L_h^k+1}^{k(i)}, \frac{\left(C^k - X_T^k\right) O_{T-L_p^k:T+L_p^k+1}^{k(i)}}{\sum_{i'=1}^{M} Q_{T-L_p^k:T+L_h^k+1}^{k(i'_k)}} \right\}$$

$$\text{for } k = 1, 2, ..., k \text{ and } i = 1, 2, ..., M) \tag{15}$$

$$I_T^{k(i)} = I_{T-1}^{k(i)} + P_T^{k(i)} - P_{T-1}^{k-1(i)} \cdot \frac{A(i, k)}{A(i, k-1)}$$

$$(\text{for } k = 1, 2, ..., K \text{ and } i = 1, 2, ..., M) \tag{16}$$

$$B_T^{0(i)} = B_{T-1}^{0(i)} + P_{T-L_h^1-1}^{(i)} - D_T^{(i)} \quad (\text{for } i = 1, 2, 3, ..., M) \tag{17}$$

$$B_T^{k-1(i)} = B_{T-1}^{k-1(i)} + P_{T-L_h^1-1}^{k-1(i)} \cdot \frac{A(i, k)}{A(i, k-1)} - P_T^{k-1(i)} \cdot \frac{A(i, k)}{A(i, k-1)}$$

$$(\text{for } k = 2, 3, ..., K \text{ and } i = 1, 2, ..., M) \tag{18}$$

4.3 Basic model analysis

Some results of an analysis of a basic model which has no restrictions on production capacity, inventory and down time are presented here to clarify the characteristics of the two types of production ordering systems.

Basic model assumptions:. Basic model of two types of production ordering systems are simplified here. The parameters of the systems are as follows;

1. $k = 9$

2. $M = 1$ and $A(1, k) = 1$ for $k = 1, 2, ..., 9$

3. $L_p^k = 1$ and $L_h^k = 0$ for $k = 1, 2, ..., 9$

4. There are irregular variations in the demand time series.

5. The forecasted demand is expressed as follows:

$$\hat{D}_{T:T+L} = \overline{D} + e_{T:T+L} \tag{19}$$

$$e_{T:T+L} = a.L.u \tag{20}$$

where, a: the degree of the forecasted error depends on the length of lead time in the forecasting period.
u :the unit disturbance depends on the normal distribution $N(0, 1^2)$.

6. The production capacity at each stage is always sufficient.

7. The amount of safety stock at each stage is always sufficient.

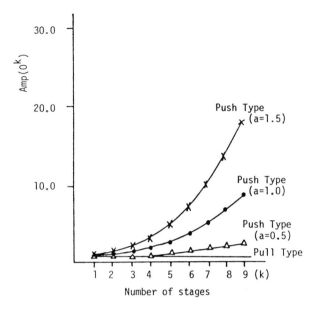

Figure 3: "Amplifications" of the ordered quantity at each stage.

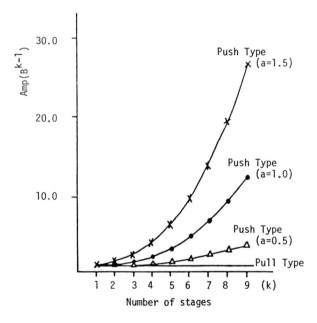

Figure 4: "Amplifications" of the inventory quantity at each stage.

8. Down time at each stage does not occur.

9. The performance measurement of the systems is defined as follows:

$$Amp(O^k) = \frac{V(O^k)}{V(D)}$$

$$Amp(B^k) = \frac{V(B^k)}{V(D)} \tag{21}$$

The "amplifications" in production ordering and inventory quantity
Figure 3 and Figure 4 show the "amplifications" in production ordering and inventory quantities on the two types of production ordering systems.

Results
    This analysis has the following conclusions:

1. In the push type production ordering system, the "amplifications" in production ordering and inventory quantities at each stage are more diffused in the stages further to final production. A larger forecasting error is likely the greater "amplifications" in production ordering and inventory quantities at the stages. Therefore, a control parameter is needed in the push type production ordering system to prevent these "amplifications".

2. In the pull type production ordering system, there is no "amplification" in the production ordering and inventory quantities. The production ordering systems without "amplifications" are effective in stabilizing the multi-stage systems and in increasing the flexibility and productivity of the systems. However, it is necessary that inventory levels are decreased at each stage in this production ordering system.

5   Conclusions

In recent years, many flexible manufacturing systems have been developed throughout the developed countries. It is important to design and operate production systems which adapt to severe changes in the market and in technology and to increase productivity.

    In this paper, the evaluative variables of flexibility are proposed and factors improving the flexibility of single stage production systems which have different characteristics are discussed. Currently, most company production systems consist of multi-stage mixed type systems. This paper also points out that the "amplification" in multi-stage production systems is one of the important factors affecting the system's flexibility and productivity. The paper also discusses some ways of reducing this "amplification" by utilizing available production ordering systems.

The multi-stage production ordering system models discussed here are simplified, and neglect the characteristics of each single production system which consist of total multi-stage production systems. The models also neglect transportation and material flow parameters. The flexible manufacturing system of today is merely treated as a stage in a succession of production stages with different characteristics; however, it should be considered as a way to increase the flexibility and productivity of the system as a whole.

References

[1] J. Hartley, FMS at work, North-Holland, Amsterdam, 1984, 286 pp.

[2] K. Takeda, S. Shimoyashiro and N. Tsuchiya, Technology and Practice of Signal System, Proceedings of 7th ICPR in Windsor, 1983, 563-566.

[3] A. Masuyama, Idea and Practice of Flexible Manufacturing System of Toyota, Proceedings of 7th ICPR in Windsor, 1983, 584-590.

[4] N. Mizoguchi, Flexible Manufacturing System for Photo-conductors, Proceedings of 7th ICPR in Windsor, 1983, pp. 591-597.

[5] J. Parnaby, Concept of a Manufacturing System, International Journal of Production Research, 17, 2 (1979) pp. 123-125.

[6] D.M. Zelenovi, c, Flexibility - A Condition for Effective Production Systems, International Journal of Production Research, 20, 3 (1980), pp. 319-337.

[7] R.E. Young, Software Control Strategies for Use in Implementing Flexible Manufacturing Systems, Industrial Engineering, 13, 11 (1981), pp. 88-96.

[8] K.E. Stecke, Loading and Control Policy for Flexible Manufacturing System, International Journal of Production Research, 19, 5 (1981) 481-490.

[9] J.A. Buzacott, The Fundamental Principles of Flexibility in Manufacturing Systems, Proceedings of the 1st International Conference on Flexible Manufacturing Systems, Brighton, UK, October (1982) 13-22.

[10] D.J. McBean, Concepts of FMS, Proceedings of the 1st International Conference on Flexible Manufacturing Systems, Brighton, UK, October (1982) 497-513.

[11] S.M. Gustavsson, Flexibility and Productivity in Complex Production Process, Proceedings of International Conference on Productivity and Quality Improvement, Tokyo, Japan, October (1982) B-3-1.

[12] P.G. Ranky, The Design and Operation of FMS - Flexible Manufacturing Systems - North-Holland, Amsterdam, 1983, p. 348.

[13] R. Muramatsu, H. Miyazaki and Y. Tanaka, Effective Production Systems which harmonized Worker's Desires with Company Needs, International Journal of Production Research, 20, 3, (1982) 297-309.

[14] T. Tabe, R. Muramatsu and Y. Tanaka, Analysis of Production Ordering Quantities and Inventory Variations in Multi-Stage Production Ordering Systems, International Journal of Production Research 18, 2 (1980) 245-257.

[15] Y. Tanaka and R. Muramatsu, A Study of the Design of a Lot Production System, International Journal of Production Research, 15, 6 (1977) 565-581.

[16] S. Kubokawa and M. Sosiroda, Scheduling Procedures for Special Order Items in a Two-stage Lot Production System, International Journal of Production Research, 18, 1 (1980) 43-56.

[17] Y. Tanaka and R. Muramatsu, An Analysis of Dynamic Characteristics of Two-stage Mixed Products Line Production Process and Lot Production Process Model, International Journal of Production Research, 20, 5 (1982) 629-641.

Part III

FMS Planning

Flexible Manufacturing Systems: Recent Developments
A. Raouf and M. Ben-Daya (Editors)
© 1995 Elsevier Science B.V. All rights reserved.

FMS Short Term Planning Problems: a Review

Mohamed Ben-Daya

Systems Engineering Department, King Fahd University of Petroleum and Minerals, Dhahran 31261, Saudi Arabia

Abstract

Flexible manufacturing systems use highly capital intensive equipment. A high system utilization must be achieved through careful short term planning. This paper deals with FMS short planning and scheduling issues. The purpose of this paper is to define and introduce the main FMS planning and scheduling problems, review the main related models and discuss the research studies dealing with these problems. Directions for future research are also outlined.

1 Introduction

Flexible Manufacturing Systems (FMS) are computer controlled batch manufacturing systems which combine the efficiency of mass production and the flexibility of job shops. An FMS consists of numerically controlled (NC) machines, automated material handling mechanisms, robots, and in-process storage facilities. An FMS operates in a large variety medium volume production environment and it is usually designed to produce a variety of high precision parts and products.

There are many complex issues associated with designing and managing an FMS. Careful planning is very important because the versatility of the NC machines gives rise to a large number of alternatives that need to be considered. Also, FMSs are highly capital intensive and therefore a high system utilization must be achieved. Once the system is in place, high system utilization can be achieved through careful short term planning.

This paper deals with FMS short planning and scheduling issues. The purpose of this paper is to define and introduce the main FMS planning and scheduling problems and discuss the research studies dealing with these problems. This is done with more emphasis on the more general methodologies and generic problems. Some directions for future research are also outlined.

This paper is intended as a first informative introduction to the important FMS short term planning problems and as a foundation to suggest useful directions for

advanced study. Our goal is to provide a basic understanding of these FMS problems
that may permit additional developments.

Before discussing the major FMS planning and scheduling problems, we need to
put these problems in perspective in the broader FMS framework. The important
issues related to FMS can be divided into five phases [41]:

1. *Design:* The design problem starts when the need for automation and flexibility
   in making parts and products is felt. The decisions involved, at this stage,
   include:

   (a) the system hardware;

   (b) computer system and control mechanisms;

   (c) FMS layout;

   (d) part families selection.

2. *Aggregate production Planning:* This phase includes the following decisions:

   (a) selection of a production planning philosophy;

   (b) a decision hierarchy for the complex FMS environment;

   (c) design and implementation of a computerized system and human/machine
       interfaces.

   The main output of aggregate production planning is a master schedule specifying part mix, production rates, and lot sizes.

3. *Short term planning or system setup:* This phase interfaces aggregate production
   planning with the day-to-day operation of an FMS. The decisions involved in this
   phase include [81]:

   (a) *Part selection problem (PSP):* The problem, at this stage, is to determine a
       subset of the set of candidate part types that are to be processed simultane-
       ously during the planning horizon. The selected set of parts must permit a
       feasible allocation of resources. The need for the PSP is due to the fact that
       the system has capacity limitation. The tool magazine capacity restricts the
       number of tools that can be mounted on the magazines and hence limits
       the number of parts that can be processed.

   (b) *Machine grouping:* This problem deals with grouping the machines of simi-
       lar types into identical machine groups. Each machine in a particular group
       is then able to perform the same operations.

   (c) *Production ratio problem:* Given the part types selection, this problem de-
       termines the relative part type mix ratios at which the selected part types
       should be produced over time.

(d) *Resource allocation problem:* This problem allocates the minimum number of pallets and fixtures of different fixture types required to maintain the production ratios found.

(e) *loading problem:* This the problem of allocating the cutting tools of all operations of the selected part types to some machine's limited capacity tool magazine.

The five short term planning problems briefly described above are linked to each other. There are different strategies for handling their interdependence in the proposed solution methods. Some authors combine many of these problems in one formulation, others disaggregate these problems and develop sequential or iterative solution procedures.

The output of this phase consists of the set of parts which will be produced during the planning period, an allocation of tools to machines, an allocation of fixtures to parts, and an assignment of unit operations to machines.

4. *Scheduling:* This phase determines the routing of parts through machine and determines start and completion times for each activity.

5. *Control:* This phase deals with the actual operation of the system. The decisions here include:

(a) design and implementation of procedures for handling machine tool and other breakdowns;

(b) periodic and preventive maintenance;

(c) quality control;

(d) on-line data collection and processing.

For an excellent review of most of these issues the reader is referred to the paper by Gunasekaran et. al. (1993) [28].

Our focus in this paper is on the important short term planning and scheduling problems arising in an FMS. Due to the high capital involved in having such a system, a high rate of efficient utilization of resources is needed to ensure an early return on investment. This can only be achieved by developing sound planning, scheduling and monitoring strategies. Stecke [81] proposed that the setup problem be divided into five subproblems as indicated earlier. However, Hwang and Shanthikumar [38] showed that only two subproblems, part selection and machine loading are important.

In this paper, we consider these two important FMS setup problems which deal with the decisions that have to be made before the FMS can start producing parts. The output of the setup phase gives the detailed machine workloads and the types of tools required for processing the selected part types to be loaded into the tool magazines on the appropriate machines. In a second stage the selected part types are scheduled through the system. Scheduling problems and their modeling and solution techniques are the object of Section 4. In section 5, we outline some directions for future research.

First, let us consider the part selection problem.

2   Part selection problem

Given the set of jobs to be processed over the short-range horizon, the part selection problem (PSP) deals with the problem of choosing a subset of parts for immediate and simultaneous processing. There are two major reasons for partitioning the jobs in an FMS into batches [14]:

1. Maintaining proper shop floor control may dictate that a smaller variety of jobs be loaded concurrently on the manufacturing facility.

2. The FMS resources are finite and limited. It is not often possible to set up the equipment with the capabilities for all required operations at one time. Such limited resources as pallets and fixtures must be changed or reassigned, and tool magazines must be reconfigured after one batch is completed so as to process another batch.

When a set of parts is selected, the number of tool slots needed to hold the tools required to process the parts should not exceed the capacity of the tool magazine. This is one of the constraints of PSP. Other constraints involve due dates and part quantities. The part quantities to be produced determine the tool processing time for each tool. This tool processing time distribution may be so unevenly distributed that there is no possibility of balancing the machine workload [35]. A common objective for PSP is to maximize the production rate. Hwang and Shogan [36] argue that this is equivalent to minimizing the number of batches. In each general form, the part selection problem not only selects part types but also part quantities. Also, due dates and other requirements have to be taken into consideration. However, taking all these factors into consideration turn the problem into an intractable one. In fact, deciding only part types is difficult enough. The problem of finding the minimum number of batches is called *the part type grouping problem* [36]. This problem can be formulated as a set covering problem.

Let us now discuss the complexity of PSP. There are three factors that make PSP difficult to solve [36]:

1. a part order has many attributes;

2. different part types may share common tools; and

3. an optimal solution requires simultaneously partitioning the production order into the minimal number of batches.

Assume that we ignore factors (1) and (2), so that only one attribute, part type, is considered and there is no tool sharing among different part types. In this simplified case, PSP is a bin packing problem which is known to be NP-Complete [24]. So, even in its simplest forms PSP is difficult. This explain the use of heuristic approaches for dealing with this problem.

There are basically two approaches for modeling and solving this setup problem:

- mathematical programming approaches; and

- group technology approaches.

2.1 Mathematical programming approaches

The following review gives a representative account of the research activity dealing with mathematical programming approaches addressing the part selection problem.

Most approaches addressing PSP consider this problem as part of a hierarchical approach dealing with many FMS planning problems. Sometimes, PSP is treated separately. PSP was one of five production planning problems identified by Stecke [81] that must be solved for efficient use of an FMS.

One of the first mathematical programming models for PSP was developed by Hwang [35]. In this paper, the author points out the drawback of the approach adopting the "similarity" concept and constructs a mathematical programming model which considers the magazine capacity constraint.

Rajagopalan (1986) [66] proposed a combined model for several setup problems including PSP. A mixed linear integer programming formulation and two types of heuristics were presented and tested on randomly generated problems. Hwang and Shanthikumar (1987) [35] discussed the production planning problem and showed that only two subproblems, part selection and machine loading, are important. They proposed a model for production planning in FMS which starts with the part selection module. They studied the effect of various part selection methods on different performance measures. Stecke and Kim (1988) [84] discussed a flexible approach to short term production planning in FMS. They demonstrated the advantages of the suggested approach via simulation. They also provided computational results of the procedures to select part types. Afentakis et al. (1989) [2] developed two heuristics for PSP for certain types of FMS. One of the proposed algorithms is based on a heuristic developed originally for bin packing (BP). The relationship between PSP and BP was discussed in the beginning of Section 2. Hwang and Shogan (1989) [36] proposed a maximal network flow model with two side constraints. The model can be relaxed to either a maximal network flow problem or two independent 0-1 knapsack problems. Sawik (1990) [72] formulated PSP, machine loading, part input sequencing, and operation scheduling as a multi-level integer program. Liang and Dutta (1992) [55] investigated the combined part selection, load sharing and machine loading problem in a hybrid manufacturing system composed of an FMS and a conventional manufacturing system. The problem is formulated as a mixed integer program.

Non-traditional optimization techniques such as simulated annealing and tabu search have been proposed recently (1993) for solving PSP [79]. This trend is likely to continue because of the success of these methods in solving several difficult problems.

2.2   Group Technology Approaches

Group technology (GT) approaches attempt to identify families of parts that require similar processing on a set of machines, and these machines are usually grouped into machine cells. Two approaches are suggested in the literature for dealing with this problem: the classification scheme using coding systems and the incident matrix approach.

The classification scheme as suggested by Kusiak [45] uses coding systems to describe the characteristic of a part based on certain attribute such as geometrical shape, types of operations required and their sequence. Most coding systems proposed use many digits to represent a part, with each digit representing an attribute. Then, clustering methods that use some measure of similarity among these codes are used to identify part families. The similarity is usually measured by the distance between two parts. A drawback of this approach is the subjectivity inherent in the coding schemes used since the attribute values are a mix of nominal and numerical data with arbitrarily chosen scales. As such, the distance between two parts depend on the coding system used [22].

Chakravarty and Shtub [10] suggested an approach which uses a part-machine incident matrix that has element $a_{ij} = 1$ if part i should be processed by machine j, and $a_{ij} = 0$, otherwise. The main idea is that if the matrix can be rearranged so that the nonzero a_{ij}'s are all clustered diagonally, the part-machine association can be easily identified. Among the algorithms proposed to find this diagonal bloc form, the rank order clustering algorithm suggested by King and Nakornchai [40] is an effective one.

Hwang [37] points out that these GT approaches fail to address adequately the differences between part families problems in GT and FMS. In FMS, one needs to take into consideration tool capacity constraints and due date interaction among parts in different batches. Also, with the enhanced variety capability, parts can be included in a family in an FMS without the same degree of processing similarity required in a GT setting.

3 Machine Loading

3.1 Introduction

When the parts to be processed simultaneously have been selected, the set of different operations that must be performed is known. The next step deals with the allocation of part operations amongst available machines for a given product mix so that some system performance criterion is optimized. This is the machine loading problem.

In a random FMS, the loading decisions are dynamic and thus reviewed periodically. The constraints of the machine loading problem include:

1. the number of tool slots available on the tool magazine of a machine spindle;

2. the number of slots a tool occupies on the magazine;

3. the nonsplitting of jobs;

4. the capacity of machines;

5. possibly some other restrictions.

Many system performance criteria have been proposed. Stecke [81] has described six objectives for the loading problem.

1. Balancing the machine processing times

2. Minimizing the number of movements

3. Balancing the workload per machine for a system of groups of pooled machines of equal sizes

4. Unbalancing the workload per machine for a system of groups of pooled machines of unequal sizes

5. Filling the tool magazine as densely as possible

6. Maximizing the sum of operations priorities

The research contributions in modeling the machine loading problem have involved both the development of models and solution procedures for solving the problem. Basically two approaches were developed to deal with the loading problem:

1. Mathematical programming approaches in which the loading problem is formulated as a mixed integer programming (MIP) problem and solved directly.

2. Given the inherent difficulty with the MIP approach, especially for large problems, various heuristics have been developed based on these formulations.

The next section reviews the literature dealing with the machine loading problem.

3.2 Machine loading literature

Although not exhaustive, the following review is representative of the research contributions developed to deal with the machine loading problem either separately or with other short term planning problems.

A leading treatment of the problem using a mathematical programming approach was developed via the 0-1 nonlinear mixed integer programming formulation of Stecke [81]. Five production planning problems are defined that must be solved for the efficient use of an FMS. The author addresses specifically the grouping and loading problems.

The relevance of the FMS loading models to the generalized transportation and assignment models has been pointed out by Kusiak [46]. A different solution methodology using an efficient branch and bound procedure has been developed by Berrada and Stecke [5]. Shanker and Srinivasulu [76] developed a two-stage branch and backtrack procedure with the objective of maximizing the assigned workload. Heuristic procedures are also developed with a bicriterion objective of minimizing the workload imbalance and maximizing the throughput for critical resources. Balasubramanian et al. [4] modeled the machine loading and tool allocation problems as a discrete generalized network with simple side constraints. They also proposed an algorithm for solving the problem. Co et al. [14] formulated the FMS batching, loading and tool configuration problems as a mixed integer program. Using submodels of the original MIP problem, they introduce a four-pass approach. The approach assumes that the need for batching is primarily that of tool magazine capacity constraints, with balancing and maximizing flexibility as secondary objectives. Stecke and Talbot [85] have suggested the need for fast heuristics that give good solutions for dealing with the loading problem. Shanker and Tzen [76] proposed a bicriterion objective of balancing the workloads among machines and meeting the job due dates for a random FMS. The heuristic methods suggested have been compared with the exact mixed integer programming solution. Using algorithms adapted from multi-dimensional bin-packing methods, Kim and Yano [39] considered the problem of assigning operations and their associated tools to machines to maximize the throughput for a specified steady-state mix of orders.

4 FMS scheduling

FMS scheduling problems are far more difficult than for production lines and job shops. Some of the reasons are as follows:

- Each machine is capable of performing many different operations.

- Several part types can be machined simultaneously.

- The system has the potential to permit alternative machine routings for a given operation giving rise to a larger number of decision variables.

- The space available for storing unfinished parts between machines is limited.

- There is a need to synchronize machines and the material handling system.

Given these differences, the methodology and techniques developed for job shops may not apply to an FMS.

The FMS scheduling problem includes scheduling machines, material handling system, and other support equipment such as pallets, fixtures and tools. The following tasks must be addressed in the FMS scheduling problem [69]:

1. Scheduling job release times

2. Sequencing the jobs and determining the start and completion times of each operation on a wide variety of resources

3. Monitoring the execution of the schedule and providing effective contingency handling

Scheduling is one of the most difficult aspects of FMS operations. The flexibility offered by such systems creates more alternatives which makes the decision process at this level more complex. Many researchers have demonstrated that scheduling decisions greatly influence FMS performance. Therefore, getting the best out of these systems requires that effective scheduling techniques be developed.

According to Rachamadugu and Stecke [65], FMS scheduling has the following important dimensions:

1. Operational mode:

   (a) Dedicated systems where small set of part types with moderate demands (i.e., 2000-200,000/year) are involved

(b) Random systems where a large set of part types with low demands (less than 2000) are involved

2. FMS type:

   (a) Flexible transfer line (flow shop type of an FMS)
   (b) General Flexible Machining Systems (job shop type of an FMS)
   (c) Flexible assembly systems (an FMS that assembles components and sub-assemblies)

3. Scheduling environment:

   (a) Static: order status changes periodically
   (b) Dynamic: order status change continuously

4. Responsiveness:

   (a) Real-time: scheduling operation by operation or by event basis
   (b) Off-line: schedule a complete set of parts at one time off-line, then the FMS manufactures these parts according to this schedule

This classification is very important in understanding and comparing studies dealing with FMS scheduling.

The literature addressing FMS scheduling issues can be divided into three categories:

- First there is research dealing with off-line scheduling whose purpose is to develop static scheduling algorithms based on analytical tools. The FMS scheduling problem is formulated as a constrained optimization model which is then solved using appropriate solution procedures.

- Second, there is literature that investigates the performance of scheduling rules using simulation models and FMS simulators and the design of on-line dispatching algorithms. The performance of machines and material handling equipment scheduling rules are tested against different scheduling criteria.

- the third category involves expert systems and/or artificial intelligence techniques. This research emphasizes the role of the expert scheduler and the importance of qualitative factors involved in FMS scheduling.

It is the purpose of this section to discuss pertinent scheduling issues and various techniques used to deal with FMS scheduling problems.

First, we present some scheduling rules that are frequently used in studying several FMS scheduling problems.

4.1 Scheduling rules

Many FMS's are operated using dispatching rules which specify which job is processed next from the queue of jobs requiring processing. A more formal definition is given below.

Definition: [92]. A dispatching rule is a function which assigns to each waiting job a scalar value, the minimum of which, among jobs waiting in the system, determines the jobs to be selected over all others for sequencing.

Scheduling rules can be static, i.e., they are used for off-line scheduling and result in a fixed schedule for the period, or dynamic, i.e., they are used for real-time scheduling and they change over time. Often, they are either based upon operations and processing or upon due dates. There is a wide base of literature available on scheduling rules. Panwalker and Iskander [63] report more than 100 rules.

In this section, we limit our review to the application of these rules in an FMS environment. The performance of a dispatching rule may be different depending on whether it is used to operate an FMS or a job shop. Therefore results from job shop studies cannot be applied directly in an FMS environment.

First, we present several scheduling rules that are quite representative of the types of rules available [59].

1. SIO (LIO) : shortest (longest) imminent time

2. SPT (LPT) : shortest (longest) processing time

3. SRPT (LRPT) : shortest (longest) remaining processing time

4. SDT (LDT) : smallest (largest) ratio obtained by dividing the processing time of the imminent operation by the total processing time for the part

5. SMT (LMT) : smallest (largest) value obtained by multiplying the processing time of the imminent operation by the total processing time for the part

6. FRO (MRO) : fewest (largest) number of remaining operations

7. FIFO : first in, first out

8. SLACK : least amount of slack (difference between due date, the present time and the remaining processing time of the part)

9. SSLACK : least amount of static slack (difference between due date, the arrival time and the total processing time of the part)

10. SLACK/RO (SSLACK/RO) : smallest ratio of (static) slack time to the number of remaining operations (slack-per-operation)

11. SLACK/TP : smallest ratio of the job slack time to the total processing time

12. SLACK/RP : smallest ratio of the job slack time to the remaining processing time

The list is by no means exhaustive. More details can be found for example in Panwalker and Iskander [63] and Montazeri and Van Wassenhove [59].

4.2 Non-traditional optimization techniques

Recently some progress has been made in solving some difficult FMS scheduling problems using non-traditional optimization techniques such as simulated annealing, tabu search, and genetic algorithms.

In the classical job shop scheduling problem there is an a priori assignment of operations to machines. This is not the case in an FMS environment where for each job a process plan is given consisting of a sequence of operations. For each operation a set of equivalent machines is available with possibly different processing times, and a joint routing and scheduling problem must be solved. A common idea used in the literature to solve this problem is based on the observation that when a routing is chosen, the joint routing and scheduling problem becomes a classical job shop problem.

In this section, we briefly discuss these methods and how they are applied to solve job shop and flow shop scheduling problems since these problems form the basis for solving an FMS scheduling problem.

4.2.1 Simulated annealing

Simulated annealing methods have gained wide attention in recent years in solving many combinatorial optimization problems. The approach has been proposed by Kirkpatrick [42]. Its basic feature is the possibility of not getting stuck in a local optimum. This is due to "hill climbing moves" controlled by some parameter. The method is less and less likely to move away from the optimal solution toward the end of the process.

Applied, for example, to the permutation flow shop scheduling problem where the measure of performance is makespan, the simulated annealing approach works as follows. Given a sequence, obtained from some other heuristic, a new sequence is generated by randomly interchanging two jobs. The new sequence is accepted if its makespan is better than that of the original sequence, otherwise it is accepted with some probability which decreases as the process evolves. The acceptance probability is of the form:

$$P_a = e^{-\alpha \Delta m},$$

where α is a control parameter which is increased during the search and Δm is the increase in makespan. If after the random pairwise interchange makespan increases, the new sequence is accepted if

$$P_a > r,$$

where r is a random number between 0 and 1. This process is terminated when a prespecified limit on the number of iterations is reached.

4.2.2 Tabu search

Tabu search was proposed in its present form by Glover [25]. It is now a well established optimization technique for solving combinatorial problems in many fields.

The basic idea of tabu search has been described in [15] as follows. Any instance of a combinatorial optimization problem is associated with a finite set of feasible solutions, each of which is characterized by a cost. The goal is to find a solution of minimum (or maximum) cost. Given a problem P, let S denote the set of feasible solutions to P and $c : S \Rightarrow \mathbf{R}$ its cost function. In order to derive a local search based algorithm for P, it is necessary to define a *neighborhood structure*, that is, a function $N : S \Rightarrow S$ which associates a set of solutions $N(s)$ with each solution $s \in S$ obtainable by a predefined partial modification of s, usually called *move*. Starting from an initial solution generated independently, a local search algorithm repeatedly replaces the current solution by a neighboring one until a superimposed stopping criterion becomes true. The algorithm returns the best solution found, with respect to the cost function.

Tabu search is a local search based optimization method; the search moves from one solution to another, choosing the best not *forbidden* element in the neighborhood. This method forbids solutions with certain attributes in order to prevent cycling and to guide the search towards unexplored regions. Without using the technique of forbidden solutions, the method can conceivably come back into the same local optimum again. However, storing complete solutions in a *forbidden list* and testing if a candidate solution belongs to the list is generally too expensive, both for memory and for computational time requirements. Usually, a *tabu list* is defined which stores only the opposite of the move applied during the search to transform a solution into a new one (i.e. the move which leads from the new solution to the old one). A solution s' is considered forbidden if the current solution s can be transformed into s' by applying one of the moves in the tabu list. In addition to a tabu status, a so-called *aspiration criterion* is associated with each move. If a current tabu move satisfies the associated aspiration criterion, it is considered an admissible move.

For a precise and complete description of this method, the interested reader can refer to the papers of Glover [25,26]. The general framework of a tabu search algorithm can be described as follows.

Procedure *TS*

begin
> <find an initial feasible solution s >;
> best := $c(s)$;
> $s^* := s$;
> Tabu-list := $\emptyset$;
> **repeat**
>> Cand(s) : $\{s' \in N(s)$: the move from s to s' does not belong to Tabu-list
>>> or it satisfies an *aspiration criterion* $\}$;
>
>> <choose $\bar{s} \in$ Cand(s) : $\bar{s}$ has the minimum *estimation* of the cost function>;
>> <put a move which leads from $\bar{s}$ to s in Tabu-list>;
>> $s := \bar{s}$;
>> if $c(s) <$ best then
>>> **begin**
>>>> $s^* := s$;
>>>> best := $c(s)$
>>> **end**
> **until** *stopping criteria* = TRUE;
> **return** s^* **end**

4.2.3 Genetic Algorithm

The idea of a genetic algorithm can be summarized as follows:

- Start with a set of initial feasible solutions which can be generated randomly. This set is called *population*. The length of the set should be an even number. Evaluate the population by evaluating each solution and storing the value of the best one.

- Generate another population by first randomly mating the solutions (i.e. each two solutions will become mates), then apply a crossover procedure for each mate. The crossover is done by randomly generating two integer numbers (I and J), I is in the range from 1 to n, and J is in the range from I to n. The operations in the positions from I to J are switched for each mate.

- A Partially Mapped Crossover (PMX) is then applied to correct any infeasible sequences (in the sense of repeating or missing operations) that might result. The idea of PMX is that, for any position inside the crossover range, if the operation that enters is already exists outside the range, then the operation outside the range will be replaced by the one that leaves from the same position.

- Apply the evaluation criteria explained before to the new population and update.

- Repeat until no substantial improvement is made.

For more details on genetic algorithm the reader is referred to the book by Goldberg [27].

4.3 Scheduling literature

4.3.1 Scheduling rules

In the following, we review the research done on evaluating scheduling rules. As we will see in this review, general results are hardly available since the performance of scheduling rules depends on the criterion chosen and, more importantly, the configuration of the FMS considered.

The performance of scheduling rules is often investigated using simulation models and FMS simulators. The following literature review is limited to studies of scheduling rules in the context of FMS.

Stecke and Solberg (1981) [80] studied an FMS consisting of ten machines, an inspection station, and a common storage area. They tested 16 rules. Their results indicate that some of the rules which are known to perform well in a conventional job shop environment performed poorly in this FMS. El-Maraghy (1982) [20] compared SIO with three other rules (FIFO, FRO, and random selection) in an FMS composed of five workstations and one load/unload station. His results indicate that SIO produced the highest production rate and the lowest mean flow time. The shortest processing time (SPT) dispatching rule has been the best performer more frequently than any other (e.g., Shanker and Tzen 1985 [75]), although no single dispatching rule clearly dominates for all criteria. Denzler and Boe (1987) [16] also performed a simulation-based experimental study in an existing dedicated FMS by using actual routing and operation times data. Their results also indicate that FMS performance is significantly affected by the choice of scheduling rules. Choi and Malstrom (1988) [13], however, performed a physical simulation to test job shop scheduling rules. Their model consisted of a miniature closed-loop. Results that differ from previous studies were presented by Co et al. (1988) [14], who investigated the effects of queue length on the relative performance of scheduling rules in a closed-loop FMS. They concluded that the scheduling decision is not likely to be as critical as in a conventional job shop.

All the studies reported above focused only on scheduling the machines rather than scheduling both the machines and the material handling system. One of the first simulation-based experimental studies which addressed the scheduling of Automated Guided Vehicles (AGV) was performed by Egbelu and Tanchoco (1984) [18]. Even though the system studied was not an FMS and machine scheduling was not directly considered, they proposed different heuristic rules for dispatching AGVs and studied their relative performance. In a later work, Egbelu (1987) [19] developed an AGV dispatching algorithm and compared it with other vehicle dispatching rules. Acree and Smith (1985) [1] also tested different cart selection and tool allocation rules. Their results indicated a significant performance difference among the tool allocation rules, but they could not find a difference in cart selection rules. Sabuncuoglu and Hommertzheim (1989) [68] developed a simulation model to test different scheduling rules. Their results indicated that scheduling AGVs is as important as scheduling machines. They also reported that the number of jobs released for processing in the system has a significant impact on the relative performance of scheduling rules. In a more recent paper (1992) [69], they used simulation to compare the mean flow time performance of different machine and AGV scheduling rules. Their results indicate that at high utilization rates, the way that machines and AGVs are scheduled can significantly affect the system performance. Szu-Yung and Wysk (1989) [92] described a scheduling algorithm which uses discrete-event simulation in combination with dynamic part dispatching rules. The algorithm combines various rules in response to the dynamic status of the system. Intuitively, by alternating rules, they will tend to compensate for the undesirable effects that each produces, and thus yield a schedule that is more sensitive to the system objectives and dynamics. Montazeri and Wassenhove (1990) [59] used a modular FMS simulator [58] to analyze the effect of some scheduling rules on the performance of a specific system. They point out that their and similar results are system dependent and should not be carelessly generalized. O'Keefe and Kasirajan (1992) [62] point out that, in an FMS, in addition to a dispatching rule, a next station selection (NSS) rule is necessary to determine the station to which the job is actually dispatched. They studied the interaction between nine dispatching and four next station selection rules in a relatively large dedicated FMS. The NSS rules investigated are:

1. NS : nearest station

2. WINQ : work in queue (select the station whose input buffer contains the smallest total amount of work, i.e., the sum of the imminent operation times for all jobs in the buffer)

3. NINQ : shortest queue (select the station with the fewest number of jobs in the buffer)

4. LUS : Lowest utilized station (select the station with the smallest total utilization rate)

Also, Choi and Malstrom [13] looked at both dispatching and NSS rules.

4.3.2 Off-line schemes

Off-line scheduling schemes include operations research techniques and various heuristics that have been proposed for dealing with the FMS scheduling problem. Most formulations available in the FMS scheduling literature consider only the scheduling of parts on the machines. This is partially due to the difficulty in formulating dynamic problems analytically.

Hitz (1979) [32] used several static problems to evaluate an off-line scheduling scheme for a dedicated, flow-shop-type system with limited buffer storage. He developed an implicit enumeration scheme to minimize the idle time of the bottleneck machine and found that the scheme yielded substantial benefits over simple real-time scheduling approaches. Sarin and Dar-El (1984) [71] investigated an FMS which operates in a dynamic environment. An off-line heuristic scheduling scheme was examined. Using a criterion of weighted machine utilization, their heuristic provided better results than the scheme the company was using. Chang et al. (1985) [10] examined a near optimal off-line scheduling heuristic for a dedicated, flow-shop FMS with a dynamic environment. They used mean flow times as the criterion and measured their heuristic against a number of real-time type schemes. They found the off-line scheme to be significantly better than the real-time schemes. However, when the real-time schemes employed look-ahead control policies, the off-line scheme was not significantly better. Yamamoto and Nof (1985) [93] studied a near optimal off-line scheduling scheme for a computer integrated manufacturing job shop facility in a static environment. Although alternative machine options were not allowed, they did include system disruption as a factor (i.e., machine failures and operational delays). In comparing their scheduling scheme against various real-time type procedures, they found that the off-line scheme performed better than the real-time schemes even when disruptions were severe. Kusiak (1986) [50] proposed a heuristic two level scheduling algorithm for a system consisting of a machining and assembly subsystem. It was shown that the upper level problem is equivalent to the two machine flow shop problem. The algorithm at the lower level schedules jobs according to established priorities. Hutchison (1988) [33] studied a near optimal off-line scheduling scheme for a random, job shop FMS within a dynamic environment. The procedure he used decomposed the problem into a loading problem and a resulting scheduling problem. This decomposed scheme was compared against seven real-time type schemes using an adjusted production rate criterion and a mean flow time criterion. The best of the real-time schemes was found to be the SPT dispatching rule coupled with a look-ahead control policy. However, the off-line scheme resulted in significantly better system performance than any of the real-time schemes. Hutchison also found that the performance of the off-line scheme, relative to the performances of the real-time schemes, improved at a faster rate as routing flexibility increased. Hutchison et. al. (1991) [34] evaluated three scheduling schemes: two off-line and one real-time scheme. The first off-line scheme solves the loading and resulting scheduling problem. The second off-line scheme decomposes the problem and solves both subproblems optimally. The real-time scheduling scheme utilizes the SPT dispatching rule coupled with a look-ahead control policy. The results indicate that both off-line schemes performed much better than the real-time scheme with respect to makespan. Also, the off-line schemes take advantage of increased routing flexibility more than the real-time scheme does. However, it is not known what would happen if disruptions and delays are taken into consideration.

It would be desirable to make all scheduling decisions simultaneously so that the best possible schedule can be found [67]. Because of computational intractability, however, hierarchical or sequential approaches are often used. Most researchers have decomposed the problem by making tool loading, part input sequencing and routing alternative operations and material handling routes a consideration in the scheduling decision. Furthermore, the dynamic nature of an FMS amplifies these problems. Raman et al. [67] presented an integer programming formulation of the problem of simultaneously scheduling jobs on machines and material handling devices. They solved the static problem as a resource constrained project scheduling problem. Alternate heuristic dispatching procedures are suggested for the dynamic case when the workloads on the machines are not balanced. Sawik (1990) [72] presented a multi-level integer programming formulation which includes the following problems: part type selection, machine loading, part input sequencing, and operation scheduling.

4.3.3 Non-traditional optimization techniques

Recently some progress has been made in solving some difficult FMS scheduling problems using non-traditional optimization techniques such as simulated annealing, tabu search, and genetic algorithms. Many papers have appeared recently dealing with the application of tabu search to solve various FMS planning problems. The list given below is by no means exhaustive.

Brandimarte et. al. (1987) [8] applied simulated annealing to solving FMS production scheduling problems. In a more recent paper, Brandimarte (1993) [7] described a hierarchical algorithm for the flexible job shop problem. A two level approach has been devised, based on the decomposition of the problem in a routing and a job shop subproblem. Both problems are tackled by tabu search. Widmer (1991) [91] used a tabu search to solve an FMS scheduling problem considering tooling constraints. A simulation experiment was carried out where problems of up to 10 jobs and 10 machines were solved using the tabu search algorithm.

The classical job shop problem has been addressed in many papers using simulated annealing (see for example [54,56]), tabu search (see for example [15,60,87], and genetic algorithm (see for example [6,23]).

4.3.4 Expert systems and artificial intelligence approaches

Many publications dealing with applications of expert systems (ES) and artificial intelligence (AI) have appeared in recent years.

Park et al [64] point out that the relative effectiveness of a given scheduling rule is likely to depend upon the system characteristics. In a dynamic system, these characteristics change over time. Therefore, it appears conceptually appealing to adopt an approach which employs appropriate and possibly different scheduling rules at various points in time. In order to do so, however, we need a mechanism which can distinguish different combinations of system characteristics from one another. AI methods are appropriate for dealing with these issues.

Today's scheduling systems, which are based on optimization techniques, misunderstand the role of the expert scheduler and underestimate the importance of qualitative factors involved in scheduling problems. Therefore scheduling problems are prime candidates for application of artificial intelligence technology. However, a more objective analysis would realize that, as optimization and expert system approaches have certain advantages in solving scheduling problems, integration of the two approaches may yield better results as discussed by Kusiak [53]. Each approach, used alone, may be more appropriate for a given problem.

For a review of expert systems in manufacturing refer to Kusiak and Chen [52].

5 Conclusions and future research directions

In this paper, three fundamental FMS short planning problems have been discussed, namely the part selection problem, the machine loading problem, and FMS scheduling problems. These problems have been introduced and pertinent literature discussed. This section offers a discussion of future research directions in these areas.

5.1 Part type selection problem

Although a lot of progress has been made in modelling and solving the part selection problem, more research is needed to include many realistic constraints that are ignored in current formulations. More realistic models should consider more attributes of part types and address the fact that different part types may share common tools, among other considerations.

5.2 Machine loading

Most of the models reported in the literature (see Section 3.2) dealing with the machine loading problem are 0-1 integer programming formulations. These models, although instrumental in understanding the problem and forming the basis for various heuristics, are very limited in addressing the problem adequately. They cannot be used directly to solve large problems. There is room for developing more efficient procedures for solving the machine loading problem. Also more attention should be given to tools, pallets, and fixtures management.

5.3 FMS Scheduling

There is a major difficulty in comparing different scheduling rules and scheduling schemes due to the lack of standardization among system definitions [28]. Results usually depend on the system configuration and cannot be generalized to other systems. Any standardization effort will be helpful in the evaluation of the solution methods developed.

Although some work has been done to compare off-line and on-line schemes for FMS scheduling, more research is needed in this direction, especially when jobs arrive randomly and disturbances to the system, such as breakdowns, are allowed.

More work is also needed in the area of synchronization of AGV's and machines. Other important issues which need further research include investigation of the effect of routing flexibility on scheduling methods.

Maintenance scheduling in an FMS environment is another topic that needs more attention from researchers [90]. As machines become more complex and automated, their availability is vital to the FMS. Efficient maintenance strategies have to be developed to prevent costly disturbances and loss of production.

Non-traditional optimization techniques such as simulated annealing, tabu search and genetic algorithms have shown a lot of promise in solving scheduling problems in general and FMS scheduling problems in particular. This trend is likely to continue.

References

[1] Acree, E.S., and Smith, M.L, (1985), "Simulation of a Flexible Manufacturing System - Applications of Computer Operating System Techniques", *The 18th Annual Simulation Symposium*, 205-216, IEEE Computer Society Press, Tampa, Florida.

[2] Afentakis, P., Solomon, M.M., and Millen, R.A., "The Part-Type Selection Problem", in: K.E. Stecke and R. Suri (eds.), *Proceedings of the Third ORSA/TIMS Conference on*

Flexible Manufacturing Systems - Operations Research Models and Applications, 1989, 141-146.

[3] Ammons, J.C., Lofgren, C.B., and McGinns, L.F., "A Large Scale Loading Problem in Flexible Assembly", *Annals of Operations Research* 3 (1985) 319-328.

[4] Balasubramanian R., Sarin, S., and Chen, C.S., "A Model and a Solution Approach for the Machine Loading and Tool Allocation Problem in a Flexible Manufacturing System," *International Journal of Production Research* **28** 637-645.

[5] Berrada, M., and Stecke, H.E., "A Branch and Bound Approach for FMS Machine Loading", in : *Proceedings of the First ORSA/TIMS Special Interest Conference on Flexible Manufacturing Systems: Operations Research and Applications*, University of Michigan, Ann Arbor MI, August 15-17, 1984.

[6] Biegel, J.E., and Davern, J.J., (1990), " Genetic Algorithms and Job Shop Scheduling", *Computers and Industrial Engineering*, **19** (1), pp. 81.

[7] Brandimarte,P., (1993), "Routing and Scheduling in Flexible Job Shop by Tabu Search", *Annals of Operations Resaerch*, **41**, pp. 157-183.

[8] Brandimarte,P., Contemo, R., and Laface, P., (1987), "FMS Production Scheduling by Simulated Annealing", *Proceedings of the 3rd on Simulation in Manufacturing*, Torino, pp. 235-245.

[9] Buzacott, J.A., and Yao, D.D., "Flexible Manufacturing Systems: A Review of Analytical Models", *Management Science* 31 (1986) 890-905.

[10] Chakravarty, A.K., and Shtub, A., "Selecting Parts and Loading Flexible Manufacturing Systems", in : *Proceedings of the First ORSA/TIMS Special Interest Confernece on Flexible Manufacturing Systems*, University of Michigan, Ann Arbor, MI, 1984, 284-289.

[11] Chang, Y.L., Sullivan, Bagchi, U., and Wilson, J., (1985), "Experimental Investigation of Real-Time Scheduling in Flexible Manufacturing Systems", *Annals of Operations Research*, **3**, pp. 355-378.

[12] Chen, I., and Chung, C.H., "Effects of Loading, and Routeing Decisions on Performance of Flexible Manufacturing Systems", *International Journal of Production Research* 29 (1991) 2209-2225.

[13] Choi, R.H., and Malstrom E.M., (1988), "Evaluation of Traditional Work Scheduling Rules in a Flexible Manufacturing System with a Physical Simulator", *Journal of Manufacturing systems*, **7** (1), pp. 33-45.

[14] Co, H.C., Jaw, T.J., and Chen, S.K., (1988), "Sequencing in Flexible Manufacturing Systems and Other Short-Queue Length Systems", *Journal of manufacturing Systems*, **7** (1), pp.1-7.

[15] Dell'Amico M., and Trubian M., (1993), "Applying Tabu Search to the Job Shop Scheduling Problem", *Annalas of Operations research*, **41**, pp. 231-252.

134

[16] Denzler, D.R., and Boe, W.J., (1987), " Experimental Investigation of Flexible Manufacturinf System Scheduling Rules", *International Journal of Production Research*, **25** (7), pp. 979-994.

[17] Doulgeri, Z., Hibberd, R.D., and Hushand, T.M., "The scheduling of Flexible Manufacturing System", *Annals of the CIRP* 36 (1987) 1-14.

[18] Egbelu, P.J., and Tanchoco, J.M.A., "Characterization of Automatic Guided Vehicle Dispatching Rules", *International Journal of Production Research* 22 (1984) 359-374.

[19] Egbelu, P.J., (1987), "Pull versus Push Strategy for Automated Guided Vehicule Load Movement in a Batch Manufacturing System", *Journal of manufacturing Systems*, **6** (3), pp. 209-221.

[20] Elmaraghy, H.A., (1982), "Simulation and Graphical Animation of Advanced Manufacturing Systems", *Journal of Manufacturing Systems*, **1**, N0. 1, pp. 53-63.

[21] El-Tamimi, A.M., Suliman, S.M.A., and Williams, D.F., (1989), "A Simulation Study of Part Sequencing in a Flexible Assembly Cell", *International Journal of Production Research* **27** 1769-1793.

[22] Eversheim, w., and Hermann, P., "Recent Trends in Flexible Automated Manufacturing", *Journal of Manufacturing Systems* 1 (1982) 139-147.

[23] Falkenauer E., and Bouffouix, S., (1991), "A Genetic Algorithm for Job Shop", *Proceedings of 1991 IEEE Conference on Robotics and Automation*, Sacramento, CA, pp. 824-829.

[24] Garetti, M., Pozzetti, A., and Bareggi, A., "On-line Loading and Dispatching in Flexible Manufacturing Systems", *International Journal of Production Research* 28 (1990) 1271-1292.

[25] Glover F., (1989), "tabu Search-Part I", *ORSA Journal on Computing*, **1**, pp. 190-206.

[26] Glover F., (1990), "Tabu Search-Part II", *ORSA Journal on Computing*, **2**, pp. 4-32.

[27] Goldberg, D.E., (1989), *Genetic Algorithms in Search, Optimization and Machine Learning*, Wiley.

[28] Gunasekaran A., Martikainen, T., and Yli-Olli, P., (1993), "Flexible Manufacturing Systems: An Investigation for Research and Applications", *European Journal of Operational Research* **66** pp. 1-26.

[29] Han, M.H., and McGinnis, L.F., "Flow Control in Flexible Manufacturing: Minimization of Stockout Cost", *International Journal of Production Research* 27 (1989) 701-715.

[30] Han, M.H., Na., Y.K., and Hogg, G.L., "Real-time Tool Control and Job Dispatching in Flexible Manufacturing Systems", *International Journal of Production Research* 27 (1989) 1257-1267.

[31] Hildebrand, R.R., "Scheduling Flexible Machine Systems when Machines are Prone to Failures", Ph.D. Thesis, MIT, Cambridge, MA, 1982.

[32] Hitz, K. (1979), "Scheduling of Flow shops. II. Report No. LIDS-R-1049, Labratory for Information and Decision Systems, MIT Cambridge, MA.

[33] Hutchison, J., (1988), "Scheduling Random Job Shop Flexible Manufacturinf Systems", Unpublished Doctoral dissertation, University of Houston, TX, USA.

[34] Hutchison, J., Leong, K., Snyder, D., and Ward, P., (1991), "Scheduling Approaches for Random Job Shop Flexible Manufacturinf Systems", *International Journal of Production Research* **29** (5), pp. 1053-1067.

[35] Hwang S., (1986), "A Constraint-Directed Method to Solve the Part Selection Problem in Flexible Manufacturing Systems Planning Stage", in : *Proceedings of the Second ORSA/TIMS Special Interest Conference on Flexible Manufacturing Systems: Operations Research Models and Applications*, pp. 297-309, Elsevier Science Publishers B.V., Amsterdam.

[36] Hwang, S.S., and Shogan, A.W., "Modeling and Solving an FMS Part Selection Problem", *International Journal of Production Research* 27 (1989) 1349-1366.

[37] Hwang, S.S., "Models for Production Planning in Flexible Manufacturing Systems", Ph.D Dissertation, University of California, Berkeley, CA, 1986.

[38] Hwang, S.S., and Shanthikumar, J.G., "An FMS Production Planning System and Evaluation of Part Selection Approaches", Management Science Working Paper No. MS-43, School of Business Administration, University of California, Berkeley, CA 1987.

[39] Kim, Y.D., and Yano, C.A. (1993), "Heuristic Approaches for Loading Problema in Flexible Manufacturing Systems", *IIE Transactions*, **25**, No.1, pp. 26-39.

[40] King, J.R., and Nakornchai, V. (1982), "Machine Component Group Formation in Group Technology: Review and Extensions", *International Journal of Production Research*, **20** (2), pp. 117-133.

[41] Kiran, A.S., (1986), "The System Setup in FMS: Concepts and Formulation", in : *Proceedings of the Second ORSA/TIMS Special Interest Conference on Flexible Manufacturing Systems: Operations Research Models and Applications*, pp. 321-332, Elsevier Science Publishers B.V., Amsterdam.

[42] Kirkpatrick, S., Gelatt, C.D., and Vecchi, M.P., (1983), "Optimization by Simulated Annealing", *Science*, **220**, pp. 621-680.

[43] Kumar, P., Singh, N., and Tewari, N.K., "A nonlinear goal programming model for the loading problem in a flexible manufacturing system", *International Journal of Production Research* 2 (1987) 13-20.

[44] Kumar, P., Tewari, N.K., and Singh, N., "Joint Consideration of Grouping and Loading Problems in a Flexible Manufacturing System" *International Journal of Production Research* 28 (1990) 1345-1356.

[45] Kusiak, A., "Part Families Selection Model for Flexible Manufacturing Systems", in: *American Industrial Engineering Conference Proceedings*, 1983, 575-580.

[46] Kusiak, A., "Flexible Manufacturing Systems: A structural Approach", *International Journal of Production Research* 23 (1985) 1057-1073.

[47] Kusiak, A., "The Part Families Problem in FMS", *Annals of Operations Research* 3 (1985) 117-124.

[48] Kusiak, A., *Modelling and Design of Flexible Manufacturing Systems*, North-Holland Amsterdam, 1986.

[49] Kusiak, A., "Application of Operational Research Models and Techniques in Flexible Manufacturing Systems", *European Journal of Operational Research* 24 (1986) 336-345.

[50] Kusiak, A., "Scheduling Flexible Machining and Assembly System", in: *Proceedings of the Second ORSA/TIMS Conference on Flexible Manufacturing Systems: Operations Research Models and applications*, pp 521-526, Elsevier Science Publishers B.V., Amsterdam.

[51] Kusiak, A., "Artificial Intelligence and Operations Research in Flexible Manufacturing Systems", *Information Systems and Operations Research* 25 (1987) 2-12.

[52] Kusiak, A., and Chen, M., "Expert Systems for Planning and Scheduling Manufacturing Systems", *European Journal of Operational Research* 34 (1988) 113-130.

[53] Kusiak, A., (1989), "Scheduling Automated Manufacturing Systems: A Knowledge-Based Approach", in: *Proceedings of the Third ORSA/TIMS Conference on Flexible Manufacturing Systems: Operations Research Models and applications*, pp 377-382, Elsevier Science Publishers B.V., Amsterdam.

[54] van Laarhoven, P.J., Aarts, E.H., and Lenstra, J.K., (1992), "Job Shop Scheduling by Simulated Annealing", *Operations Research*, **40** (1), pp. 113.

[55] Liang, M., and Dutta, S.P., "A Mixed Integer Programming Approach to the Machine Loading and Process Planning Problem in a Process Layout Environment", *International Journal of Production Research* 28 (1990) 1471-1484.

[56] Lo, Z.P., and Bavarian, B., (1992), "Optimization of Job Shop Scheduling on Parallel Machines by Simulated Annealing Algorithms", *Expert Systems with applications*, **4** (3), pp. 323.

[57] van Looveren, A.J., Gelders, L.F., and Van Wassenhove, L.N., "A Review of FMS Planning Models", in : A. Kusiak (ed.), *Modelling and Design of Flexible Manufacturing Systems*, Elsevier Science Publishers, Amsterdam, 1986, 3-31.

[58] Montazeri, M., Gelders, L.F., and van Wassenhove, L.N., (1988), "A Modular Simulator for Design, Planning, and Control of Flexible manufacturing Systems", *The International journal of Advanced Manufacturing Technology*, **3** (1), pp. 15-32.

[59] Montazeri, M., and Van Wassenhove, L.N., "Analysis of Scheduling Rules for an FMS", *International Journal of Production Research* 28 (1990) 785-802.

[60] Mooney, E.L., and Rardin, R.L., (1993), "Tabu Search for a Class of Scheduling Problems", *Annals of Operations Research*, **41**, pp. 253.

[61] O'Grady, P.J., and Menon, U., "Loading a Flexible Manufacturing System", *International Journal of Production Research* 25 (1987) 1053-1068.

[62] O'Keefe, R.M., and Kasirajan, T., (1992), "Interction Between Dispatching and Next Station Selection Rules in a dedicated Flexible manufacturing System", *International Journal of Production Research* **30** (8), pp 1753-1772.

[63] Panwalker S.S., and Iskander W. (1987), "A Survey of Scheduling Rules", *Operations Research,* **25**, pp. 45-61.

[64] Park. S., Raman, N., and Shaw, M.J., "Heuristic learning for Pattern-Directed Scheduling in a Flexible Manufacturing System", in: K.E. Stecke and R. Suri (eds.), *Proceedings of the Third ORSA/TIMS Conference on Flexible Manufacturing Systems - Operations Research Models and Applications* 1989, 133-139.

[65] Rachamadugu, R. and Stecke K.E. (1987), "Classification and Review of FMS Scheduling Procedures", Working paper # 481R, Graduate School of Business Administration, University of Michigan, Ann Arbor, MI.

[66] Rajagopalan, S., (1986), "Formulation and Heuristic Solutions for Parts Grouping and Tool Loading in Flexible Manufacturing Systems", in : *Proceedings of the Second ORSA/TIMS Special Interest Conference on Flexible Manufacturing Systems: Operations Research Models and Applications*, pp. 311-320, Elsevier Science Publishers B.V., Amsterdam.

[67] Raman, N., Talbot, B., and Rachamadugu, R., "Simultaneous Scheduling of Material Handling Devices in Automated Manufacturing", in : K.E. Stecke and R. Suri (eds.), *Proceedings of the Second ORSA/TIMS Conference on FMS: Operational Research Models and Applications*, Elsevier Science Publishers, Amsterdam, 1986.

[68] Sabuncouglu, I., and Hommertzheim, D.L., "An Investigation of Machine and AGV Scheduling Rules in an FMS", in: K.E. Stecke and R. Suri (eds.), *Proceedings of the Third ORSA/TIMS Conference on Flexible Manufacturing Systems - Operations Research Models and Applications*, 1989, 79-84.

[69] Sabuncouglu, I., and Hommertzheim, D.L., (1992), "Experimental Investigation of FMS machine and AGV Scheduling Rules Against the Mean Flow-Time Criterion", *International Journal of Production Research* **30** (7) pp. 1617-1635.

[70] Sarin, S.C., and Chen, S.C., "The Machine Loading and Tool Allocation Problem in a Flexible Manufacturing System", *International Journal of Production Research* 25 (1987) 1081-1094.

[71] Sarin, S.C., and Dar-El, E. (1984), "Approaches to the Scheduling Problems in Flexible Manufacturing Systems", *The Industrial Engineering Conference Proceedings*, Institute of Industrial Engineers.

[72] Sawik, T., "Modelling and Scheduling of a Flexible Manufacturing System", *European Journal of Operational Research* **45** (1990) 177-190.

[73] Seidmann, A., and Schweitzer, P.J., "Part Selection Policy for a Flexible Manufacturing Cell Feeding Several Production Lines", *IIE Transactions* 16 (1984) 355-362.

[74] Shalev-Oren, S., Seidmann, A., and Schweitzer, P.J., "Analysis of Flexible Manufacturing Systems with Priority Scheduling: PMVA", *Annals of Operations Research* 6 (1985) 115-139.

[75] Shankar, K., and Tzen, Y.J., J., "A Loading and Dispatching Problem in a Random Flexible Manufacturing System", *International Journal of Production Research* 23, (1985) 579-595.

[76] Shankar, K., and Srinivasalu, A., "Some Solution Methodologies for Loading Problems in a Flexible Manufacturing System", *International Journal of Production Research* 27 (1989) 1019-1034.

[77] Shaw, M.J., "A Pattern-Directed Approach to FMS Scheduling", *Annals of Operations Research* 15 (1988) 353-376.

[78] Solot, P., "A Concept for Planning and Scheduling in an FMS", *European Journal of Operational Research* 45 (1990) 85-95.

[79] B. Srivastava and W.H. Chen, "Part Type Selection Problem in Flexible Manufacturing Systems: Tabu Search Algorithms", *Annals of Operations Research*, **41**, (1993), pp. 279-297.

[80] Stecke, K.E. and Solberg, J.J., Loading and Control Polcies for a Flexible Manufacturing System", *International Journal of Production Research* 19 (1981) 481-490.

[81] Stecke, K., "Formulation and Solution of Non-linear Integer Production Planning Problem for Flexible Manufacturing Systems", *Management Science* 29 (1983) 272-285.

[82] Stecke, K.E., "Design planning Scheduling and Control Problems of Flexible Manufacturing Systems", *Annals of Operations Research* 3 (1985) 3-12.

[83] Stecke, K.E., and Solberg, J.J., "The Optimality of the Unbalanced Workloads and Machine Group Sized for FMS", *Operations Research* 33 (1985) 882-910.

[84] Stecke, K., and Kim, I., (1986), "A Flexible Approach to Implementing the Short Term FMS Planning Function", in : *Proceedings of the Second ORSA/TIMS Special Interest Conference on Flexible Manufacturing Systems: Operations Research Models and Applications*, pp. 283-295, Elsevier Science Publishers B.V., Amsterdam.

[85] Stecke, K., and Talbot, B., "Heuristic Loading Algorithms for Flexible Manufacturing Systems", Working Paper No. 348, Division of Research, Graduate School of Business Admiistration, The University of Michigan, Ann Arbor, MI, 1986.

[86] Stecke, K.E., "A Hierarchical Approach to Solving Machine Grouping and Loading Problems of Flexible Manufacturing Systems", *European Journal of Operational Research* 24 (1986) 369-378.

[87] Taillard, E. (1989), "Parallel Tabu Search Technique for the Job Shop Scheduling Problem", Research Report ORWP 89/11, Ecole Polytecnique Federale de Lausanne.

[88] Tang, C.S., "A Job Scheduling Model for a Flexible Manufacturing Machine", in: *1986 International Conference on Robotica and Automation*, 1 (1986) 152-155.

[89] Whitney, C.K., and Suri, R., "Algorithms for Part and Machine Selection in Flexible Manufacturing Systems", *Annals of Operations Research* 3 (1985) 34-45.

[90] Widmer, M., and Solot, P., (1990), "Do not Forget the Breakdowns and the Maintenance Operations in FMS Design Problems", *International Journal of Production Research* **28** pp. 421-430.

[91] Widmer, M., (1991), "Job Sop Scheduling with Tooling Constraints: a Tabu Search Approach", *Journal of the Operational Society*, **42**, pp. 75-82.

[92] Wu, S.Y.D., and Wysk, R.A., "An Application of Discrete-Event Simulation to On-Line Control and Scheduling in Flexible Manufacturing", *International Journal of Production Research* 27 (1989) 1603-1623.

[93] Yamamoto, M. and Nof, S., (1985), "Scheduling/Rescheduling in the Manufacturing Operating System Environment", *International Journal of Production Research* **23** (4), pp. 705-722.

[94] Yao, D.D., and Buzacott, J.A., "Modelling a Class of State-Dependent Routeing in Flexible Manufacturing Systems", *Annals of Operations Research* 3 (1985) 153-167.

[95] Yao, D.D., and Buzacott, J.A., "Models of Flexible Manufacturing Systems with Limited Local Buffers", *International Journal of Production Research* 24 (1986) 107-118.

[96] Yih, Y. and Thesen, A., "Semi-Markov Decision Models for Real-Time Scheduling", *International Journal of Production Research* 29 (1991) 2331-2346.

Flexible Manufacturing Systems: Recent Developments
A. Raouf and M. Ben-Daya (Editors)
1995 Elsevier Science B.V.

Loading Models In Flexible Manufacturing Systems

A. Kusiak

Department of Industrial Engineering,4132 University Building
The University of Iowa, Iowa City, IA 52242, USA

Abstract

In this paper two different approaches to planning and scheduling in the classical manufacturing systems are overviewed. Based on these approaches, a planning and scheduling methodology for Flexible Manufacturing Systems (FMSs) is presented. One of the most important problems in the FMS methodology, the machine loading problem, is discussed to a great extent. Four models of the loading problem are formulated. Some of the algorithms for solving the loading problems are also discussed.

1   Introduction

Flexible Manufacturing Systems (FMSs) have gained a large number of applications. To date hundreds of these systems have been implemented around the world. Japan is leading in the FMSs international competition in terms of the number of applications and associated management and organization successes. It would be difficult to further develop FMSs without knowing their nature.

The aim of this paper is to discuss a new FMS planning and scheduling methodology. Based upon the framework of this methodology, a number of loading models are formulated.

Planning and scheduling problems are very difficult to solve computationally (see Lenstra 1977, VanWassenhove and DeBodt 1980). This is probably the main reason why a relatively small number of planning and scheduling theoretical results have been applied in practice. Taking into account on one hand the limitation of theory, and on the other the need for optimization of planning and scheduling decisions in FMSs, perhaps the best remedy is to reduce the size of these problems. This can be done by aggregation and decomposition.

It is not intended in this paper to survey all the planning and scheduling models encountered in the classical manufacturing systems, but to show only typical approaches. Readers interested in the classical planning and scheduling models may refer, for example, to the excellent surveys written by Gelders and VanWassenhove (1981), Graves (1981), Elmaghraby (1978), and Silver (1981).

2 Planning and Scheduling in the Classical Manufacturing Systems

There are two basic approaches to production planning and scheduling in the classical manufacturing systems:

1. material requirements to planning (MRP)

2. hierarchical production planning (HPP)

2.1 Material Requirement Planning

To date, Material Requirement Planning (MRP) systems have been widely applied in industry. We should talk about a class of MRP systems rather than a single MRP approach, because of their diversity (Graves 1981). A typical MRP system is presented in Figure 1.

Based on product requirements which are either deterministic or forecasted orders, a master schedule is generated. The master schedule specifies the number of products to be manufactured with each time period (i.e. 1 month) over a planning horizon (i.e. 1 year). For each time period, products are exploded to parts. Of course, different products may consist of common parts, which are summarized. Production expressed in parts is split into batches. The formulae currently being applied for calculating the batch sizes are diverse. In general, they try to incorporate inventory holding costs and set-up costs.

Part batches are then assigned (loaded) to the most appropriate group of machines, then scheduled to ensure high machine rate utilization and satisfaction of the imposed constraints, i.e. due dates. The decision processes behind the MRP system are of an iterative nature. They are usually repeated many times until acceptable machine loads and sequence of batches are generated.

2.2 Hierarchical Production Planning

Hierarchical production planning systems take advantage of a natural approach for solving complex problems. Firstly, the production planning is solved on aggregated data over a long planning horizon. As a result of this, the entire production planning problem is decomposed into subproblems which are computationally easier to handle. Depending on the nature of the overall production planning problem, there may be two or more decision levels in the structure of the hierarchical planning system.

There are the following two advantages to an aggregate approach as opposed to a detailed planning approach (Bitram and Hax 1977):

- aggregate demands can be forecast more accurately than their desegregate components

- there is a reduction in a problem size and degree of data detail.

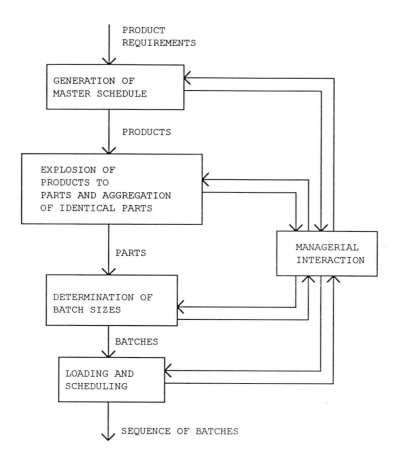

Figure 1: Decision blocks in a typical MRP system

The hierarchical methodology presented in this paragraph is based on the deterministic approach presented by Bitran, Haas and Hax (1981). Dempster (1982), and Dempster et al. (1981) have discussed a stochastic approach to the HPP systems.

Before the hierarchical production planning approach is outlined the following terms are defined (see Bitran et al. 1981):

product	–	an item delivered to the customer
product type	–	a group of products having similar production costs and inventory holding costs.
product family	–	a group of items pertaining to the same product type and sharing similar setups.

The structure of the hierarchical production planning system is shown in Figure 2.

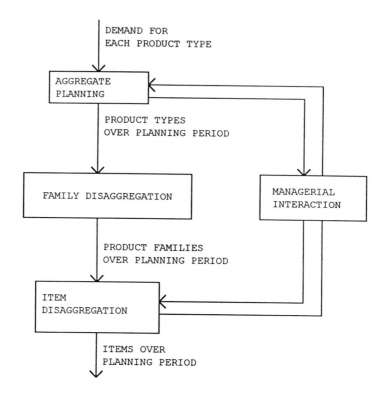

Figure 2: Hierarchical Planning System

Mathematical formulations of each of the three decision blocks in Figure 2 are discussed in Bitran et al. (1981), Bitran and Hax (1977), and Bitran et al.(1982).

3   Planning and Scheduling Methodology Applicable to FMSs

Most literature on FMSs has been published in the last decade, and there have not been many successful applications of planning and scheduling methodologies in the classical manufacturing systems; therefore, it would be difficult to expect an efficient methodology applied to FMSs.

Two recent survey papers, one by Buzacott and Yao (1982) and Sarin and Wilhelm (1983), are devoted entirely to the modeling of FMSs. It should be emphasized that most of the literature published on FMSs is based on queuing theory. An integer programming approach to the problems of FMSs has been underestimated. It seems that the integer programming approach embedded in an efficient information system can be used to model and efficiently solve many of the planning and scheduling problems, including the dynamic problems. Such an information system is discussed in Suri and Whitney (1984).

Based on the classical planning and scheduling approaches discussed in this paper, a methodology applicable to FMSs is outlined. The proposed methodology tries to incorporate optimization into the planning and scheduling decision process to a large extent. The problems being subject to optimization are of reduced sizes, either through aggregation or decomposition. Such an approach assures solution of these problems within a practically acceptable time. The new FMS planning and scheduling methodology is presented in Figure 3.

The aggregate planning (level 1 in Figure 3) is, to some degree, similar to the aggregate model of Hax and Meal (1982). To reduce the problem size, the products are grouped into product types. It is necessary, however, to modify Hax and Meal's (1975) model to make it applicable to FMSs. The modification includes both the objective function and the constraints. First of all, the workforce factor can be eliminated in the aggregate model of Hax and Meal (1975). A typical objective function in an FMS aggregate model would include the sum of inventory costs, production costs and costs representing production capacity. It should be stressed that the production capacity can be a decision variable in the aggregated FMS planning model.

In the classical systems the production capacity was varied typically by hiring and firing policies. In FMSs, it is possible to vary the production capacity only by changing the technological parameters; i.e. machining speed or feed rate. It should be remembered that this also influences the production costs.

Resource grouping (level 2 in Figure 3) is a very important issue in FMSs and is discussed mainly in the parts context in Kusiak (1984c). Grouping of parts and machines enables decomposition of the overall planning problems into subproblems. As indicated in Kusiak (1984c), grouping of machines into Flexible Manufacturing Cells (FMCs) is considered as a logical grouping opposite to the physical cellular concept in the classical manufacturing systems.

After the parts and the machines have been grouped, a new problem of allocating the part families to FMCs arises. This problem is solved in level 3 in Figure 3. In the classical Group Technology (GT) concept, parts have been grouped into families. Typically one part family would be assigned to one manufacturing cell. As mentioned previously, in an FMS process of grouping of machines into FMCs is of a dynamic nature. The modularity and flexibility of the materials handling system enables logical grouping of machines according to the current production needs.

Grouping of machines in FMSs becomes a software issue, rather than the hardware layout problem in the GT manufacturing concept. In the planning and scheduling methodology proposed in this paper, the part families concept is also different than in the classical GT systems. We should look at generation of part families as an aggregation process. At this stage (level 2 in Figure 3) parts can be aggregated subject

146

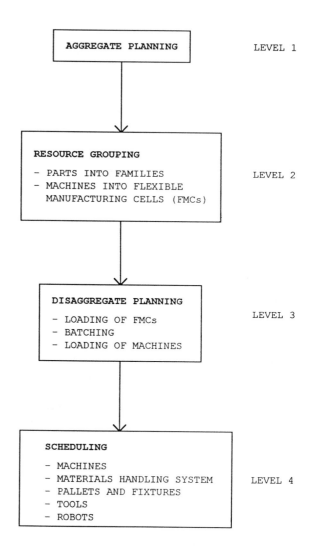

Figure 3: FMS planning and scheduling methodology

to similarity of

- tools

- fixtures

- pallets

- robot grippers

- machines

A number of different models could be formulated, however, the exact considerations would be beyond the scope of this paper. It should be stressed that this methodology assumes a relatively large number of families. On average, there will be more than one part family loaded to one FMC in level 3 in Figure 3. The objective function of the FMCs loading may have many different forms. As an example, the minimization of production costs or travel costs could be considered. The model's constraints could be concerned with the number of part families loaded on an FMC, number of tools, number of pallets, etc.

In level 3, after the FMC loading problem has been solved, it may be necessary to split parts into batches. This, however, depends on the length of the planning period of level 1 and order sizes. The short planning period and the small size of orders may result in small batch sizes. If this would not occur, the batching model related to the aggregated model of Bitran et al. (1981) could be formulated for part families.

After the FMCs have been loaded with part families, another optimization problem arises within each FMC. There is always some flexibility in assigning operations to machines. The problem of loading of machines with operations (level 3 in Figure 3) can be modeled in a number of different ways. Loading problems of this nature have been formulated in Kimemia and Gerschwin (1980), Kusiak (1984b) and Stecke (1983).

One of the advantages of the proposed methodology is that the resulting scheduling problems (level 4 in Figure 3) can become relatively easy to solve. The loading performed in level 3 may decompose the multi-machine scheduling problem, for example into a one- or two-machine scheduling problem. The number of machines involved in the resulting scheduling problem depends on the loading problem formulation (see Kusiak 1984b). As stated in Kusiak (1984a), there are other machine scheduling issues which would be considered in FMSs; namely, scheduling of:

- materials handling system

- pallets and fixtures

- tools

- robots

4 Machine Loading Problem

One of the most important problems in the presented FMS planning and scheduling methodology is that of machine loading (see level 3 in Figure 3). There is extensive literature on the loading problem. One of the most recent surveys on this subject is presented in Stecke and Solberg (1981).

The classical loading problem falls into a class of problems known in the operations research literature as the assignment problem (Ross and Soland 1980), the generalized transportation problem (Pogany 1978), the bin packing problem (Jonson 1974), the multiple knapsack problem (Hung and Fisk 1978), and the loading problem (Deane and Moodie 1972, Christofides, Mingozzi and Toth 1979).

Let us concentrate on two of the above mentioned formulations, namely:

- the generalized assignment problem

- the generalized transportation problem

Given appropriate interpretation to these two operations research problems; they become loading problems in the manufacturing systems. In manufacturing systems, typically, the batches of parts are being loaded. Of course, there are a number of operations to be performed on each part. In this paper, for the conceptual and the notational convenience loading of batches of operations is considered.

4.1 Loading Model Based on the Generalized Assignment Problem

This loading model is based on a formulation of the generalized assignment problem. In order to present this model let the following notations be introduced:

I set of batches of operations to be processed

J set of stations

T_{ij} time of processing batch i on station j, for each $i \epsilon I$ and $j \epsilon J$

C_{ij} cost of processing batch i on station j, for each $i \epsilon I$ and $j \epsilon J$

b_j processing time available on station j, for each $j \epsilon J$

$y_{ij} = \begin{cases} 1 & \text{if batch } i \text{ is processed on station } j \\ 0 & \text{otherwise} \end{cases}$

The objective is to minimize the total sum of the processing costs,

$$\textbf{(MA)} \qquad \text{minimize} \qquad \sum_{i \epsilon I} \sum_{j \epsilon J} C_{ij} y_{ij} \qquad (1)$$

$$\text{subject to:} \qquad\qquad\qquad\qquad\qquad\qquad (2)$$

$$\sum_{j \epsilon J} y_{ij} = 1, \quad \text{for each } i \epsilon I \qquad (3)$$

$$\sum_{i \epsilon I} T_{ij} y_{ij} \leq b_j \;\;, \quad \text{for each } j \epsilon J \qquad (4)$$

$$y_{ij} = 0, 1; \quad \text{for each } i \epsilon I \text{ and } j \epsilon J \qquad (5)$$

Constraints (2) ensure each batch of operations is processed on exactly one station. Constraints (3) ensure each station availability time cannot be exceeded.

One of the most effective algorithms for solving model MA has been developed by Ross, Soland and Zolteners (1980). Their algorithm is of the branch and bound type, with one of the bounds based on Lagrangian relaxation. They reported on computational results for many generalized assignment problems.

As an example consider the problem for $m = 30$ and $n = 40$ which was solved in 0.207 seconds, where m is number of stations and n is number of batches.

4.2   Loading Model Based on the Generalized Transportation Problem

Consider the assigning of a set I of batches of operations to a set J of stations. Each of these stations is capable of processing any batch of operations with different efficiency.

Let us denote:

t_{ij} unit time of processing an operation from batch i on station j, for each $i \epsilon I$ and $j \epsilon J$

c_{ij} unit cost of processing an operation from batch i on station j, for each $i \epsilon I$ and $j \epsilon J$

a_i required number of operations from batch i, for each $i \epsilon I$

b_j processing time available on station j, for each $j \epsilon J$

x_{ij} number of operations from batch i to be processed on station j, for each $i \epsilon I$ and $j \epsilon J$.

The loading model to MT can be formulated as follows:

$$\textbf{(MT)} \qquad \text{minimize} \qquad \sum_{i \epsilon I} \sum_{j \epsilon J} c_{ij} x_{ij} \tag{6}$$

$$\text{subject to:} \tag{7}$$

$$\sum_{j \epsilon J} x_{ij} = a_i, \quad \text{for each } i \epsilon I \tag{8}$$

$$\sum_{i \epsilon I} t_{ij} x_{ij} \le b_j \quad, \quad \text{for each } j \epsilon J \tag{9}$$

$$x_{ij} \ge 0, \quad \text{for each } i \epsilon I \text{ and } j \epsilon J \tag{10}$$

Constraints (6) specify that the required number of operations from batch i is equal to a_i. Constraint (7) ensure that the time available on each station j is not exceeded. There has been only a few attempts to solve the model MT reported in the literature. Charnes and Cooper (1954) presented the stepping stone algorithm for solving model (MT), which has been generalized by Eisemann (1964).

4.3 Machine Loading Models Applicable to FMSs

Based on the previous considerations we will present four new machine loading models applicable to FMSs. In formulation of these models we will try to stick to the following two principles:

- formulate the models in a simple way

- use the practical assumptions

Loading Model M1

One of the requirements in many of the FMSs is that each operation could be processed on more than one station. The loading model M1 presented below incorporated this requirement.

To formulate model M1, in addition to notation introduced in model MT, define the following parameters:

d_{ij} maximum allowable time for processing set of operations from batch i on station j, for all $i \epsilon I$ and $j \epsilon J$

n_i maximum number of stations the set of operations from batch i can be processed on, for each $i \epsilon I$.

The formulation of model M1 is as follows:

$$\text{(M1)} \qquad \text{minimize} \qquad \sum_{i \epsilon I} \sum_{j \epsilon J} c_{ij} x_{ij} \qquad (11)$$

$$\text{subject to:} \qquad\qquad\qquad\qquad\qquad\qquad (12)$$

$$\sum_{j \epsilon J} x_{ij} = a_i, \quad \text{for each } i \epsilon I \qquad (13)$$

$$t_{ij} x_{ij} \le d_{ij} y_{ij}, \quad \text{for each } i \epsilon I \text{ and } j \epsilon J \qquad (14)$$

$$\sum_{j \epsilon J} y_{ij} \le n_i, \quad \text{for each } i \epsilon I \qquad (15)$$

$$x_{ij} \ge 0, \quad \text{integer}, \quad \text{for each } i \epsilon I \text{ and } j \epsilon J \qquad (16)$$

$$y_{ij} = 0, 1, \quad \text{for each } i \epsilon I \text{ and } j \epsilon J \qquad (17)$$

Loading Model M2

There are constraints which are usually of no concern to management of the classical manufacturing systems, but which play an important role in FMSs. These constraints are concerned with a limited tool magazine capacity. In order to formulate a model which will incorporate a tool limit constraint, let us introduce the following parameters:

k_{ij} space occupied by the tool required for manufacturing of operations from batch i on station j, for each $i \epsilon I$ and $j \epsilon J$

f_j station j tool magazine capacity, for each $j \epsilon J$.

Formulation of the loading model M2 is as follows:

$$\textbf{(M2)} \qquad \text{minimize} \qquad \sum_{i \epsilon I} \sum_{j \epsilon J} c_{ij} x_{ij} \qquad (18)$$

subject to: $\qquad\qquad\qquad\qquad\qquad\qquad\qquad\qquad\qquad$ (19)

$$\sum_{j \epsilon J} x_{ij} = a_i, \quad \text{for each } i \epsilon I \qquad (20)$$

$$t_{ij} x_{ij} \leq d_{ij} y_{ij}, \quad \text{for each } i \epsilon I \text{ and } j \epsilon J \qquad (21)$$

$$\sum_{i \epsilon I} k_{ij} y_{ij} \leq f_j, \quad \text{for each } j \epsilon J \qquad (22)$$

$$\sum_{j \epsilon J} y_{ij} \leq n_i, \quad \text{for each } i \epsilon I \qquad (23)$$

$$x_{ij} \geq 0, \text{ integer, } \quad \text{for each } i \epsilon I \text{ and } j \epsilon J \qquad (24)$$

$$y_{ij} = 0, 1, \quad \text{for each } i \epsilon I \text{ and } j \epsilon J \qquad (25)$$

Loading Model M3

Apart from the constraints introduced in models M1 and M2, there are also constraints associated with the length of tool life. To formulate loading model M3, in addition to the notation in model M2, let us denote

r_{ij} expected length of life of a tool applied in processing operations
from batch i on station j

$$\textbf{(M3)} \qquad \text{minimize} \qquad \sum_{i \epsilon I} \sum_{j \epsilon J} c_{ij} x_{ij} \qquad (26)$$

subject to: $\qquad\qquad\qquad\qquad\qquad\qquad\qquad\qquad\qquad$ (27)

$$\sum_{j \epsilon J} x_{ij} = a_i, \quad \text{for each } i \epsilon I \qquad (28)$$

$$\sum_{i \epsilon I} t_{ij} x_{ij} \leq b_j, \quad \text{for each } j \epsilon J \qquad (29)$$

$$t_{ij} x_{ij} \leq r_{ij} y_{ij}, \quad \text{for each } i \epsilon I \text{ and } j \epsilon J \qquad (30)$$

$$\sum_{j \epsilon J} y_{ij} \leq n_i, \quad \text{for each } i \epsilon I \qquad (31)$$

$$x_{ij} \geq 0, \text{ integer, } \quad \text{for each } i \epsilon I \text{ and } j \epsilon J \qquad (32)$$

$$y_{ij} = 0, 1, \quad \text{for each } i \epsilon I \text{ and } j \epsilon J \qquad (33)$$

Loading Model M4

This model incorporated features of model M2 and M3

$$(\text{M4}) \qquad \text{minimize} \qquad \sum_{i \epsilon I} \sum_{j \epsilon J} c_{ij} x_{ij} \qquad\qquad (34)$$

subject to: $\qquad\qquad\qquad\qquad\qquad\qquad\qquad\qquad (35)$

$$\sum_{j \epsilon J} x_{ij} = a_i, \quad \text{for each } i \epsilon I \qquad\qquad (36)$$

$$\sum_{i \epsilon I} t_{ij} x_{ij} \leq b_j, \qquad \text{for each } j \epsilon J \qquad\qquad (37)$$

$$\sum_{i \epsilon I} k_{ij} y_{ij} \leq f_j, \quad \text{for each } j \epsilon J \qquad\qquad (38)$$

$$t_{ij} x_{ij} \leq r_{ij} y_{ij}, \quad \text{for each } i \epsilon I \text{ and } j \epsilon J \qquad (39)$$

$$\sum_{j \epsilon J} y_{ij} \leq n_i, \quad \text{for each } i \epsilon I \qquad\qquad (40)$$

$$x_{ij} \geq 0, \quad \text{integer}, \quad \text{for each } i \epsilon I \text{ and } j \epsilon J \qquad (41)$$

$$y_{ij} = 0, 1, \quad \text{for each } i \epsilon I \text{ and } j \epsilon J \qquad (42)$$

5 Solving The FMS Loading Models

The presented FMS loading models (M1-M4) belong to a class of mixed integer (integer and boolean) programming problems. They can be solved by a general mixed integer programming techniques, i.e. cutting plane and branch and bound. In practice, however, it may be difficult to obtain optimal solutions to the FMS loading models in an acceptable time, due to their large size (for example m = 20, n = 2000). Taking into account the FMS environment, in many cases it is satisfactory to generate a feasible solution. Computational experience with models related to M1-M4 indicates that such a feasible solution, in fact, may be very close to the optimal one. This is mainly due to the data structure associated with the existing FMSs; namely, the cost coefficients c_{ij} in the objective function are typically uniformly distributed.

These is also one more FMS feature which should be mentioned, namely the planning horizon corresponding to the loading models. The models discussed (M1-M4) are static FMS models, but they can be applied in dynamic situations as well. In the static applications (long planning horizon, for example 24 hours), the time to solve these models is not a crucial factor. One can imagine a situation where the long horizon results become invalid, because of FMS disturbances; i.e. machine failure. This generates a dynamic situation which requires recomputation of the loading model. In this case the length of the computing time becomes a very crucial factor. In such a case, one might be satisfied with the suboptimal solution.

Taking into account the features of the FMS environment discussed, a class of subgradient algorithms seems well suited for solving the four FMS loading models. These algorithms, typically generate a feasible solution in a modest computing time.

It may take a considerably longer time to find an optimal solution; or, in many cases to confirm the optimality of the previously found feasible solution. A class of subgradient algorithms also has an advantage of indicating the quality of any feasible solution found. A relative distance of a feasible solution from the optimal one is very often used as a stopping criterion.

To illustrate the efficiency of a subgradient algorithm, consider the following formulation of the segregated storage problem.

$$\textbf{(PS)} \quad \text{minimize} \quad \sum_{i \in I} \sum_{j \in J} c_{ij} x_{ij} \tag{43}$$

$$\text{subject to:} \tag{44}$$

$$\sum_{j \in J} x_{ij} = a_i \quad , \quad \text{for each } i \in I \tag{45}$$

$$x_{ij} \leq b_j y_{ij} \quad , \quad \text{for all } i \in I \text{ and } j \in J \tag{46}$$

$$\sum_{i \in I} y_{ij} \leq 1 \quad , \quad \text{for all } j \in J \tag{47}$$

$$x_{ij} \geq 0, \text{integer}, \text{for each } i \in I \text{ and } j \in J \tag{48}$$

$$y_{ij} = 0, 1 \quad , \text{for each } i \in I \text{ and } j \in J \tag{49}$$

This problem is a special case of the loading model M1.

The computational results of solving the problem (PS) on the CDC CYBER 170/720 by the subgradient algorithm are demonstrated in Table 1.

Table 1. Computational results

Problem size		Average Number of Iterations	Average Computing Time in CPU sec.
m	n		
20	10	20	0.50
25	20	9	0.84
25	25	15	1.29
30	20	4	0.88
40	30	16	2.9

The average number of iterations and the average computing time in Table 1 are reported for optimal solutions to the problem PS. A more detailed computational analysis is presented in Gunn and Kusiak (1983).

6 Conclusions

In the first part of this paper the planning and scheduling methodologies of the classical manufacturing systems were discussed. As these methodologies do not satisfy the needs of FMSs, a new planning and scheduling concept was outlined. This concept should fill the existing gap in the theory and applications of FMSs. It may also indicate the future research directions towards more efficient control strategies in FMSs.

The second part of this paper shows a relationship between FMS machine loading models and some of the known operations research models. Four new formulations of the FMS loading problem are presented. The strength of these formulations is in their linear structure. There are good prospects for applications of these models in the existing FMSs. This is mainly due to the constraints encompassing a large variety of industrial situations.

Acknowledgement

This research was partially supported by the Natural Sciences and Engineering Research Council of Canada.

References

[1] G.R. Bitran, E.A. Hass and A.C. Hax, "Hierarchical production planning system: A two-stage system", *Operations Research*, **30**, (1982), 232-251.

[2] G.R. Bitran, E.A. Hass and A.C. Hax, "Hierarchical production planning: A single stage system", *Operations Research* , **29**, (1981), 717-743.

[3] G.R. Bitran and A.C. Hax, "On the design of hierarchical production planning systems", *Decision Sciences*, **8**, (1977), 28-55.

[4] J.A. Buzacott and D.D. Yao, "Flexible manufacturing systems: A review of models", Working Paper No. 82-007, Department of Industrial Engineering, University of Toronto, (1982).

[5] A. Charnes and W.W. Cooper, "The stepping stone method of explaining linear programming calculations in transportation problems", *Management Science*, (October 1954) 49-69.

[6] N. Christofides, A. Mingozzi and P. Toth, "Loading problems", in Christofides, N. et al. (Eds.) *Combinatorial Optimization*, Wiley, New York (1979) 425 pp.

[7] R.H. Deane and C.L. Moodie, "A dispatching methodology for balancing workload assignments in job shop production facility", *AIIE Transactions*, **4**, (1972) 277-283.

[8] M.A.H. Dempster, "A stochastic approach to hierarchical planning and scheduling", in Dempster, M.A.H. et al. (Eds.) **Deterministic and Stochastic Scheduling**, D. Reidel Publishing Company, Dordrecht, Holland, 1982, pp. 419.

[9] M.A.H. Dempster, M.L. Fisher, B. Legewang, L. Janssen, J.K. Lenstra and A.G.H Rinnooy Kan, "Analytical evaluation of hierarchical planning systems", *Operations Research* , **29**, (1981), 707-717.

[10] K. Eisemann, "The generalized stepping stone method for the machine loading model", *Management Science*, **11**, (1964), 154-176.

[11] S.E. Elmaghraby, "The economic lot scheduling problem (ELSP): Review and extensions", *Management Science*, **24**, (1978), 587-598.

[12] L.F. Gelders and L.N. Van Wassenhove, "Production Planning: A review", *European Journal of Operational Research*, **7**, (1981), 101-110.

[13] E.A. Gunn and A. Kusiak, "A methodology for application of Fenchel's duality theory to large scale problems". Working Paper No. 05/82, Dept. of Industrial Engineering, Technical University of Nova Scotia, Canada, (1982).

[14] S. Graves, "A review of production scheduling", *Operations Research*, **29**, (1981), 646-675.

[15] A.C. Hax and H.C. Meal, "Hierarchical integration of production planning and scheduling", in Geisler, M.A. Ed., *TIMS Studies in Management Sciences, 1, Logistics*, North-Holland, Amsterdam, (1975), 53-69.

[16] M.S. Hung and J.C. Fisk, "An algorithm for 0-1 multiple kanpsack problems", *Naval Research Log. Quarterly*, **25** (1978), 571-579.

[17] J.G. Kimemia and S.B. Gerschwin, "Multicommodity network flow optimization in flexible manufacturing systems", Report No. ESL-FR-834-2, Electronic Systems Laboratory, MIT, Cambridge, (1980).

[18] A. Kusiak, "Analysis of flexible manufacturing systems", Working Paper # 09/83, Department of Industrial Engineering, Technical University of Nova Scotia, Halifax, Nova Scotia, (1983).

[19] A. Kusiak, "Flexible manufacturing systems: A structural approach", Working Paper No. 4/84, Department of Industrial Engineering, Technical University of Nova Scotia, Halifax, Nova Scotia, (1984a).

[20] A. Kusiak, "Loading models in flexible manufacturing systems", *Proceedings of the 7th International Conference on Production Research*, Windsor, Ontario, (1984b), 641-647.

[21] A. Kusiak, "The part families problem in flexible manufacturing systems", working paper No. 06/84, Department of Industrial Engineering, Technical University of Nova Scotia, Halifax, Nova Scotia, (1984c).

[22] D. Jonson, "Fast algorithm for bin packing", *Journal of Computer and System Science*, **8**, (1974), 272-314.

[23] J.K. Lenstra, "Sequencing by enumerative methods", Mathematical Center Tract, Mathematisch Centrum at Amsterdam, (1977).

[24] Z. Pogany, "An algorithm for solving the generalized transportation problem", *Proceedings of the 8th IFIP Conference on Optimization Techniques*, Wurzburg, September 5-9, (1977), Springer Verlag, New York (1978).

[25] G.T. Ross and R.M. Soland, "A branch and bound algorithm for the generalized assignment problem", *Mathematical Programming*, **18**, (1975) 91-103.

[26] G.T. Ross, R.M. Soland and a.A. Zolteners, "The bounded interval generalized assignment model", *Naval Research Logistics Quarterly*, **27**, (1980), pp. 625-633.

[27] K.E. Stecke and J.J. Solberg, "The optimal Planning of Computerized Systems, The CMS Loading Problem", Report No. 20, School of Industrial Engineering, Purdue University, (1981).

[28] S.C. Sarin and W.E. Wilhelm, "Models for the design of flexible manufacturing systems", *Proceedings of the Annual IIE Conference*, Louisville, KY., (1983), 564-574.

[29] E.A. Silver, "Operations research in inventory management: A review and critique", *Operations Research*, **29**, (1981), 628-645.

[30] K.E. Stecke, "Formulation and solution of nonlinear integer production planning problems for flexible manufacturing systems", *Management Science*, 29, (1983), 273-288.

[31] R. Suri and C.K. Whitney, "Decision support requirements in flexible manufacturing", *Journal of Manufacturing Systems*, **3**, (1984).

[32] L.N. VanWassenhove and M.A. DeBodt, "Capacitated lot sizing for injection molding: A case study", Working paper # 80-26, Katholieke Universiteit Leuven, (1980).

Flexible Manufacturing Systems: Recent Developments
A. Raouf and M. Ben-Daya (Editors)
© 1995 Elsevier Science B.V. All rights reserved.

Production Planning Model for a Flexible Manufacturing System

Pankaj Chandra

Faculty of Management, McGill University
1001 Sherbrooke St. W., Montreal, Quebec, Canada H3A 1G5

Abstract

In this paper we develop a production planning model which considers the problem of selecting optimal routes for manufacturing various part types in a Flexible Manufacturing System. We address the planning problem and its shop floor implications in a closed loop framework thereby making order release policies more effective. An hierarchical solution procedure is outlined. We illustrate the model with an example.

1   Introduction

Short product life cycles, mass customization, and short delivery times are some of the important challenges that industries are facing today. Firms are looking for practices and technologies that will help them in addressing such needs. Recent advances in manufacturing technology, which cater to the growing needs of the industry, have strengthened the case for improving or introducing flexibility in manufacturing systems. Flexible manufacturing system (FMS) is such an example of this new technology. It comprises workcenters (consisting of machines) which are connected by a transportation system and are capable of performing a number of different tasks – all under computer control, thereby achieving improved productivity and reduction in the costs with losing sensitivity to product quality. The machines have automatic tool changing capabilities which enable them to perform a set of operations while minimizing set up delays. Several different types of parts can be processed on a flexible manufacturing system. Parts, in general, have the flexibility to follow a variety of workcenter sequences and are made to demand (i.e., pull system). This reduces the lot size, sometimes to as low as one. The purpose of this paper is to provide an analytical methodology for production planning in a flexible manufacturing system.

FMSs can be analyzed using queuing network models for performance evaluation and optimization for system design and planning (Chatterjee, et. al., 1984) [5]. In their model, a workpiece goes through a series of operations at various workcenters before it is unloaded from the system. A path for a given workpiece is defined as a sequential set of operations. The flexibility of the system allows for various kinds of parts to be processed at the same time. The workcenter may consist of a single machine capable of performing a single job or multiple jobs (i.e., multi-tool machine with automated tool changing capabilities) or a group of machines. Hence, a workpiece which enters the system has the option of going to several different workstations. This is governed by the type of operation required on a job and gives rise to at least one path which the job may follow before being unloaded. Therefore, the network of workcenters and the transportation system (connecting workcenters and the load/unload center) allows many possible paths for a workpiece to move through the system. It has been found that variety and fluctuations in the demand of various workpieces or parts being routed influence the nature of the FMS routing procedure (Jaikumar, 1984) [6] and, consequently, the design of the planning system.

The pioneering work in the area of analytical modelling of FMS was done by Solberg (1978)[14] with the development of the CAN-Q model based on a closed queuing network. Kimemia and Gershwin (1978)[8] have presented an optimization model with workcenter level complexity (i.e., with all machines at a workcenter having the same process time for all part types) for an open-queuing network. Secco Suardo (1978)[13] coupled a non-linear programming formulation with a closed-queuing network model for a single class of jobs.

Stecke (1983)[15] has defined a set of production planning problems and attempts to solve the pooling and the loading subproblems. It has been suggested that work load should remain unbalanced when the size of machine groupings (i.e., after maximum possible pooling) is unequal (Stecke, 1983)[15] . Wittrock (1985)[20] used heuristic algorithms to solve the part mix allocation problem, the objective being minimization of makespan or total completion time of each day's mix. He argues for studying setup times since, often, part types are grouped into families and a machine incurs a setup time while shifting from one family to another. Needless to say, there is a need to carry out capacity planning and operations scheduling before hand in order to guarantee a high usage rate of the expensive flexible manufacturing system (Schaluch, 1982)[12]. He observed that planning results would become better in measure as the degree of freedom for planning, such as alternative machines, alternative sequence of operations, and alternative operation variants are taken into consideration. Stecke and Kim (1988)[18] have presented a model which selects a subset of part types to be machined together and determines their aggregate production ratios over some upcoming period. The authors use CAN-Q for determining the relative ratios of workloads on machine types.

Several authors (including Stecke and Solberg, 1981[15]; Ammons et. al., 1984[1]; Chakravarty and Shtub, 1984[3]; Rajagopalan, 1986 [11]) have addressed various issues related to loading and control of FMSs. None, however, incorporate the state of the shop in their formulations. Few attempts have been made to solve the production planning and scheduling problems simultaneously. Kusiak (1986)[9] , Stecke (1986)[17] and Nelson (1986)[10] propose hierarchical production planning frameworks which involve solving the planning and scheduling problems in a top down fashion. To our best knowledge, this remains as the most feasible approach available. The results of Karmarkar (1987)[7] show the need for incorporating the shop floor status in determining order release policies and consequently planning decisions. Kusiak (1986)[9] outlines the necessity of exploiting the "interaction between FMS and its environment". One way of making production planning more reactive, in this hierarchical framework, is to address the planning problem and the shop floor implications in a closed-loop framework. This would render order releases to a scheduling module more relevant and effective.

Given a fixed mix of parts (as defined by its demand) and knowing the workcenters that can perform each operation, we want to determine the optimal number of parts to be processed on each feasible route and subsequently the workcenters on which to process them such that the overall production rate of the system is maximized or the total variable costs are minimized. Such a planning exercise becomes necessary in light of the above discussion, especially, since it forms the basis for production scheduling.

In the next section we develop a mathematical programming model which maximizes the throughput rate of the system while the build up of queues within the system is modelled as a constraint. An illustrative example is provided in Section 3 and conclusions are presented in Section 4. Details of the solution procedures are given in the Appendix.

2 Optimization Model for a FMS and the Solution Algorithm

In this section we develop a production planning model for a flexible manufacturing system. We also describe a solution algorithm for solving the above problem. The manufacturing system comprises a set of workcenters which is a collection of machines with a dedicated inter-machine transport facility. Each machine can perform one or more operations. One or more operations is required to process any given part type. A route is defined as a sequence of workcenters (and machines) which a part type visits before it becomes an end product.

Given a product mix, we seek to determine the mix of routes which maximize the total production rate for all the different part types in the manufacturing system. The parameters and variables that describe the problem are:

Parameters

p = number of part types
d = number of operations
n = number of workcenters
b = number of machine types
r_i = number of feasible routes for producing part type i
S_{ml} = setup time for operation m on machine type l
O_{imlj} = processing time for part type i for operation m on machine l at workcenter j
d_i = demand for part type i
C_l = available capacity at machine l

a_i = fraction of part type i in the total production

I = maximum allowable work-in-process inventory

$t_{(j,k)}$ = travel time between workcenters j and k

Variables

X_{iljr} = production rate of part type i on machine type l at workcenter j on route r

PR_i production rate of part type i over all routes

$q_j(X_{1j}...X_{pj})$ = average queue length at workcenter j comprising different part types

The optimization problem PP which maximizes the overall throughput rate of the system subject to the constraint imposed on the average level of work-in-process inventory can be formulated as follows:

$$\textbf{PP}: \text{Maximize} \quad \sum_{i=1}^{p} PR_i$$

$$\text{subject to:}$$

$$PR_i = \sum_{r=1}^{r_i} X_{il^*j^*r},$$
$$\forall j^* \in M(j^*), \ \forall l^* \in M(l^*j^*) \quad (1)$$

$$PR_i = a_i . \sum_{i=1}^{p} PR_i \quad \forall i \quad (2)$$

$$\sum_{j=1}^{n} q_j(X_{ij}...X_{pj}) + \sum_{i=1}^{p}\sum_{r=1}^{r_i}\sum_{l=1}^{b} (X_{iljr}.t_{(j,k)}) \leq I \quad \forall(j,k) \in A \quad (3)$$

$$\sum_{r} X_{il^*j^*r} \geq d_i \quad \forall i \quad (4)$$

$$\sum_{i}\sum_{m}\sum_{r} X_{iljr}.O_{imlj} \leq C_l \quad \forall j,l \quad (5)$$

$$X_{iljr} \text{ and } PR_i \geq 0 \quad \forall i,j,l,r \quad (6)$$

Constraint 1 establishes the relationship between the production rate for each part type and the production rates at bottleneck stations on each route for every part type. The production rate on any route, for a given part type, is given by the production rate of the bottleneck machine on that route. Since it is possible that a machine can be on more than one route, the workload at any given machine comprises amount produced on different routes and, may be, consisting of different part types. It can also be appreciated that the bottleneck workcenter may be different for each route. l^* is the bottleneck machine on route r for part type i and is determined using the algorithm outlined in the Appendix. Similarly, j^* is the workcenter which includes the bottleneck machine l^* on the route r. $M(j^*)$ is the set of bottleneck workcenters and $M(l^*j^*)$ is the set of bottleneck machines in each workcenter, j^*. Constraint 2 is the ratio requirement for each part type i such that $\sum_i a_i = 1$. It can capture the relationship between part types as defined by the bill of material. Constraint 3 ensures that the average WIP does not exceed the prescribed level. The WIP level is the sum of average queue lengths at all workcenters and the pipeline stock (i.e., stock under transportation between workcenters). A is the set of all workcenter pairs which are connected by the material handling system. The average queue length at any workcenter includes arrivals from other workcenters (or other machines in the same workcenter) as well as external arrivals. However, if $j = 1$ is the load station and $j = n$ is the unload station, then q_1 = external queue while all other workcenters have no external queues. Constraint 4 ensures that the production rate over all routes, for each part type, meets the demand for that part type. Constraint 5 ensures that the amount produced on any given machine, over all possible routes, does not exceed the capacity of the machine. Constraint 6 are the non-negativity constraints.

The above mentioned production planning model embodies a spectrum of issues ranging from the determination of routes for each part type to the determination of average queue lengths. We have developed an hierarchical solution algorithm (which is shown in Figure 1) for solving the problem PP. The heuristic procedure involves solving an LP along with a queuing subproblem PP. The heuristic procedure involves solving an LP along with a queuing subproblem - these two phases are tied together in a closed-loop framework. As shown in Figure 1, Phase I determines the proportion of a given part type that should be manufactured on each of the feasible routes (i.e., the starting solution); this subsequently forms the input to queue-length-determination problem which provides us with the average queue length at each workcenter. Phase II uses the average queue values from Phase I to form the production planning problem. It determines the production quantity for each part type on every feasible route. Each phase provides input to the other - the cycling process, thereby, updates and improves the optimal part-route mix at the end of each iteration until it converges. The hierarchical solution procedure requires enumeration of feasible routes for each part type in terms of operation and machine visits. Any standard route generation algorithm can be used for obtaining the routes (e.g., route generator of Chatterjee et. al., 1984[5] or one can be developed using the constructs in Chandra, 1993 [4]). The initialization of the solution procedure is done via an aggregate determination of the quantity of given part types that should be manufactured on a given feasible route (i.e., β_{ir}). This has been dealt as SUBPROBLEM 2 in the Appendix. SUBPROBLEM 1 in the Appendix outlines a procedure for determining the bottleneck machine on each feasible route and discusses related issues. The queuing subproblem (i.e., the determination of the average queue length at each workcenter) is solved using the following heuristic which modifies Suri and Hilderbrant's (1984)[19] variant of Schweitzer-Bard algorithm for the above problem:

$$W_{ij} = \left\{ \left[\frac{\sum_{r=1,j\in R}^{r_i} \beta_{ir} - 1}{\sum_{r=1,j\in R}^{r_i} \beta_{ir}} \right] . Q_{ij} . O_{imlj} \right\} + \left\{ \sum_{v=1,v\neq i}^{p} Q_{vj} . O_{vmlj} \right\}$$

W_{ij} is the average waiting time for part type i at workcenter j and Q_{ij} is the average queue length at workcenter j as part type i enters the workcenter. All other variables

162

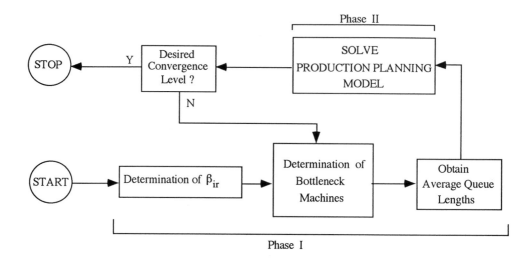

Figure 1: Hierarchical Solution Procedure

have their defined meaning. The above heuristic is employed to determine $q_j(X_{1j}...X_{pj})$ using 'Mean Value Analysis' (Suri and Hilderbrant, 1984)[19] for the multiple product case. The heuristic converges in finite number of steps (convergence results are shown in Section 3).

The structure of the general LP optimization problem (i.e., PP) is such that a solution procedure can be developed by application of the Dantzig-Wolfe price directive decomposition procedure for solving the LP problem for the entire network (especially for large size problems as described in Chandra (1993) [4]). For small size problems one can use any standard simplex code (e.g., LINDO).

3 Numerical Results for a Four Station System

Consider the system depicted in Figure 2. There are two workcenters in the system and the load and unload stations. There are two types of parts being manufactured. Each workcenter comprises two machine types. Each workcenter is connected by a dedicated conveyor. Three different kinds of operations can be performed by the four machines. Operation 1 needs to be performed on part type 1 while operation 2 and 3 need to be performed on part type 2.

$d = 3 \rightarrow [m = 1, 2, 3]$
$n = 4 \rightarrow [j = L, 1, 2, UL]$
$p = 2 \rightarrow [i = 1, 2,]$
$b = 4 \rightarrow [l = L1, L2, L3, L4]$

- Load and Unload operations take 1 minute each

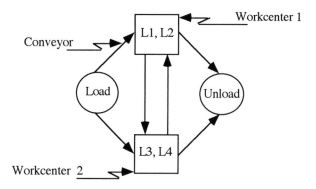

Figure 2: A Four Station System

- Average travel time on each arc = 1 minute, i.e., $t_{(j,k)} = 1$ min. for all $(j, k) \in A$

- $a_1 = a_2 = 0.5$

- The average operation times (in minutes) on each machine type are given as follows:

Operation	Machine Type			
	1	2	3	4
1	12	-	-	11
2	8	7	-	-
3	-	-	13	-

Routings for the Two Part Types

Part type 1 -

        ROUTE 1 : L - W1(L1) - UL
        ROUTE 2 : L - W2(L4) - UL

Part tyope 2 -

        ROUTE 3 : L - W1(L1) - W2(L3) - UL
        ROUTE 4 : L - W1(L2) - W2(L3) - UL
        ROUTE 5 : L - W2(L3) - W1(L1) - UL
        ROUTE 6 : L - W2(L3) - W1(L2) - UL

where W1 and W2 are workcenters 1 and 2 and L1, L2, L3 and L4 are the machines used at those workcenters. Since this design is simple, routes can be enumerated by inspection. Alternatively, for a complex design, the route generation algorithm of Chatterjee et. al. (1984)[5] can be employed for this task using the constructs and matrices that have been developed in Chandra (1993)[4].

The starting solutions (β_{ir}) for the above routes were determined using Subproblem 2. From the analysis of bottleneck machines (Subproblem 1) for the above routes for part type 1, the bottleneck machines are L1 for Route 1 and L4 for Route 2. For part type 2, the bottleneck machine is L3 over all routes. These form the starting solution to the overall problem as well as inputs to the queuing subproblem.

Queuing Subproblem

The computer implementation for determining the average queue length (using the modified Schwitzer-Bard heuristic given in the earlier section; this is used within the mean value analysis framework) was done in Fortran 77. For the base total demand of 120 units for the entire planning period, the results obtained for the four workcenter system, were as follows:

Average Queue Length at Node 1 : 4
Node 2 : 4
Node 3 : 4
Node 4 : 4

The heuristic was tested for convergence and it was found to converge in a finite number of steps. The following table exhibits the number of iterations taken to converge, the cpu time used on DEC 10, and the average queue lengths obtained for a ten workcenter FMS while varying the part types from 2 to 10:

Table 1: Performance of the queueing heuristic.

No. of Part Types	No. of Workcenters	No. of Iterations	CPU Secs.	Total No. of Parts in Queue
2	10	6	0.47	20
4	10	7	0.91	40
6	10	7	1.17	50
8	10	7	1.45	70
10	10	7	2.01	92

The efficacy of this implementation vis-a-vis other methods (which needless to mention are also approximations) still remains an exercise.

Now that all the subproblems have been solved the original production planning problem (PP) is formulated and solution obtained. The value of I is the limit on work-in-process inventory in the system for a given demand for each part type. Total number of parts in queue in the system is 16. Results were obtained for the current demand level (i.e., 60 units of each part type). Subsequently, the demand level was varied and the whole exercise repeated. Movement of parts from one workcenter to another with the increase in demand levels was studied.

At the base demand levels (60 units of each part type per planning period) all of type 1 parts followed route 2, i.e., L - W2(L4) - UL. The proportion of part type 1 that use workcenter 2 is shown in Figure 3 as a function of the work-in-process inventory.

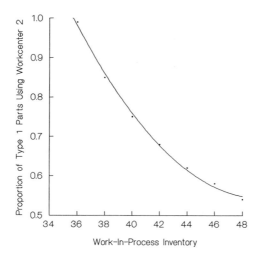

Figure 3: Workcenter 2 Usage by Part Type 1

When the in-process inventory is low the proportion of part type 1 using workcenter 2 is high since it is the faster workcenter (operation time for operation 1 is lowest at machine L4). As the number of parts in the system increases, more type 1 parts start utilizing workcenter 1. Similar results were reported by Secco-Suardo (1978)[13] while optimizing a closed network of queues. However, in his study the proportion of parts visiting the faster station depended on the ready number of pallets. Figure 4 shows the variation of the production rate of part type 1 with I and was found to be directly proportional to the work in process inventory. Nevertheless, maximum production rate stabilizes at 10.45 parts/hr. as I tends to infinity. At that production rate both the machines (L1 and L2) which can perform operation 1 on part type 1 are utilized to the fullest.

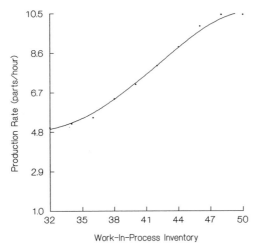

Figure 4. Production rate levels.

166

Type 2 parts, on the other hand, always follow route 4: L - W1(L2) - W2(L3) - UL. They go to workcenter 1 first and then to workcenter 2. Workcenter 2 (machine L3) is the bottleneck hence using route 6(L - W3 (L3)- W1(L2) - UL) increases the work-in-process inventory correspondingly.

4 Conclusions

We have presented an optimization model for production planning in a Flexible Manufacturing System. Our model takes into consideration the state of the shop in making decisions regarding the quantity of different products that are to be produced on various routes in the manufacturing system. We develop an hierarchical solution procedure that solves the planning and the queuing subproblems in an iterative fashion to obtain the results. An example was solved to exhibit the feasibility of the proposed system and results leading to the convergence of the queuing heuristic were reported. We also discuss the implications of the results that were obtained. We feel that this closed-loop framework at an early stage of production planning makes the plan more reactive and effective when implemented via the scheduling process as it considers the impact of loading on shop floor queues while making the production planning decisions.

This exercise of determining the optimal proportion of each part type to be manufactured on the various feasible routes is important from the point of view of the scheduling problem. Development of a scheduling heuristic for this system and a study of the performance of the hierarchical planning and scheduling problem would be the next step in this research.

Acknowledgment

This research has been partially supported by NSERC Research Grant OGP004250. Valuable comments by Professor Morris Cohen and Christopher Jones and an ananymous referee are gratefully acknowledged.

References

[1] Ammons, J.C., C.B. Lofgren, and L.F. McGinnis (1984) "A Large-Scale Workstation Loading Problem", in K.E. Stecke and R. Suri (eds.), *Proceedings of the First ORSA/TIMS Conference on Flexible Manufacturing Systems*, Ann Arbor, pp. 249-255.

[2] Buzacott, J.A. and D.D. Yao, (1986) "Flexible Manufacturing Systems: A Review of Models" *Management Science*, 32, pp. 890-905.

[3] Chakravarty, A.K. and A. Shtub, (1984) "Selecting Parts and Loading Flexible Manufacturing Systems", in K.E. Stecke and R. Suri (eds.) *Proceedings of the First ORSA/TIMS Conference on Flexible Manufacturing Systems*, Ann Arbor, pp. 284-289.

[4] Chandra, P., (1993) "Production Planning Model for a Flexible Manufacturing System", Working Paper, Faculty of Management, McGill University, Montreal.

[5] Chatterjee, A., M.A. Cohen, and W. Maxwell, (1984) "Manufacturing Flexibility: Models and Measurements", in K.E. Stecke and R. Suri (eds.) *Proceedings of the First ORSA/TIMS Conference on Flexible Manufacturing Systems*, Ann Arbor, pp. 49-64.

[6] Jaikumar, R. (1984) "Flexible Manufacturing Systems: A Managerial Perspective", Working Paper # 1-784-078, Harvard Business School.

[7] Karmarkar, U.S., (1987) "Lot Sizes, Lead Time and In-Process Inventories", *Management Science*, 33, pp. 409-418.

[8] Kimemia, J.G. and S.B. Gershwin, (1978) "Network Flow Optimization in a Flexible Manufacturing System", *IEEE Conference on Decision and Control*, pp. 633-639.

[9] Kusiak, A., (1986) "Application of Operational Research Models and Techniques in Flexible Manufacturing Systems", *European Journal of Operational Research*, 14, pp. 336-345.

[10] Nelson, C.A., (1986) "A Mathematical Programming Formulation of Elements of Manufacturing Strategy: FMS Applications", in K.E. Stecke and R. Suri (eds.) *Proceedings of the Second ORSA/TIMS Conference on Flexible Manufacturing Systems*, Elsevier Science Publishers, B.V. Amsterdam, pp. 31-42.

[11] Rajagopalan, S. (1986) "Formulation and Heuristic Solutions for Part Grouping and Tool Loading in Flexible Manufacturing Systems", in K.E. Stecke and R. Suri (eds.) *Proceedings of Second ORSA/TIMS Confernece on Flexible Manufacturing Systems*, Elsevier Science Publishers, B.V. Amsterdam, pp. 311-320.

[12] Schlauch, R., (1982) "Production Scheduling in FMS", Technical Report, Frauhofer Institute for Manufacturing Engineering & Automation, Stuttgart, W. Germany.

[13] Secco Suardo, G. (1978) "Optimization of a Closed Network of Queues", *LIDS-MIT* Report No. ESL - FR-834-3, M.I.T., Cambridge.

[14] Solberg, J.J. (1978) "Analytical Performance Evaluation for the Design of Flexible Manufacturing Systems", *Proceedings of the IEEE Decision and Control Conference*, San Diego, CA, pp. 640-646.

[15] Stecke, K.E. and J.J. Solberg, (1981) "Loading and Control Policies for a Flexible Manufacturing System", *International Journal of Production Research*, 19, pp. 481-490.

168

[16] Stecke, K.E. (1983) "Formulation and Solution of Non-Linear Production Planning Problems for Flexible Manufacturing Systems", *Management Science*, , 29, pp. 273-288.

[17] Stecke, K.E. (1986) "A Hierarchical Approach to solving Machine Grouping and Loading problems of Flexible Manufacturing Systems", *European Journal of Operational Research*, 24, pp. 369-378.

[18] Stecke, K.E. and I. Kim, (1988) "A Study of FMS Part Type Selection Applications for Short Term Production Planning, *International Journal of Flexible Manufacturing System*, 1, pp. 7-29.

[19] Suri, R. and R.R. Hildebrant, (1984) "Modeling Flexible Manufacturing Systems Using Mean Value Analysis", *Journal of Manufacturing Systems*, 3, pp. 27-38.

[20] Wittrock, R.J. (1985) "Scheduling Algorithms for Flexible Flow Lines", *I.B.M. Journal of Research & Development*, 29, pp. 401-412.

APPENDIX

Subproblem 1 : Determination of the Bottleneck Machine

Let β_{ir} = units of part type i on route r; and c^i_{ljr} = the load on machine l at workcenter j on a given route r for a given part type i (we shall currently drop the superscript i since the analysis is for every part type).

Load can be expressed as

$$c_{ljr} = \left[\left(\beta_{ir} \cdot \sum_m O_{imlj} \right) + \sum_m S_{ml} \right]$$

STEP 1

Calculate c_{ljr} for each workcenter on every feasible route for a given part type i.

STEP 2

Determine if the machine l falls on a single route. In case a machine falls on two or more routes, then it's effective load is the sum of the c_{ljr} for all the routes which use machine l at workcenter j; e.g., if machine l falls on routes 1 and 2; then

$$c^e = c_{lj1} + c_{lj2}$$

where c^e = effective load.

STEP 3

Choose the workcenter and machine with the highest c^e value as the bottleneck.

EXAMPLE

Consider a simple network showing two routes for a part type (Figure 5). The boxes are workcenters (wk) and figures above them are operation times (assume setup = 0). Then at:

wk1 : $c^e = 5(\beta_1 + \beta_2)$
wk2 : $c^e = 4\beta_1$
wk3 : $c^e = 7\beta_2$
wk4 : $c^e = 6(\beta_1 + \beta_2)$

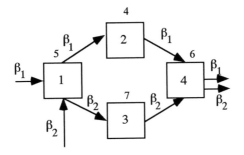

Figure 5. Flows between workcenters.

Hence, bottleneck for the route = max c^e over routes 1, 2, 3, & 4. Clearly, bottleneck shall change as β^s change.

Subproblem 2 : Determination of β

Let β_{ir} = units of part type i manufactured on route r. β_{ir} must be determined since it forms the input to the queuing subproblem. The approach is to model all the feasible routes into a network and then solve the min-time flow problem with n workcenters. The objective is to determine β_{ir} such that the flow minimizes the sum of operations and move times.

Let $G = (X, U)$ be a directed graph (assumed to be connected). Associated with each arc u(i.e., (j, k)) is an overall time value, τ_u^i which is the sum of the operation and move time required by a part type on any given route.

A is the totally unimodular, vertex-arc-incidence matrix of graph G such that

$$A = a_{ju}, j = 1, ..., N \text{ and } u = 1, ..., M$$

where each column corresponds to an arc of G and each row to a vertex of G, then we can say that for each row $j(j\epsilon X)$

$$w^+(j) = \{u|a_{iu} = 1\}$$
$$w^-(j) = \{u|a_{iv} = -1\}$$

and for each column u,

$$a_{ju} = 1 \text{ and } a_{ku} = -1 \text{ if } u = (j, k) \in U$$

We can say that β_i is a vector of M elements, i.e.,

$$\beta_i = \left[\beta_i^1, ..., \beta_i^u, ..., \beta_i^M\right] \in R^M$$

where

$$\beta_i^u = \sum_{r=1, u=(j,k), k\epsilon R} \beta_{ir} \quad (R = \text{route matrix for } r)$$

or in other words, the effective flow on an arc is the sum of flows of part types i on all routes which use workcenters j and k. Now we can write the conservation of flow at each workcenter on the route

$$\sum_{u\epsilon w^+(j)} \beta_i^u = \sum_{u\epsilon w^-(j)} \beta_i^u \quad \forall j \in X \qquad (I)$$

which can also be written in terms of the vertex-arc incidence matrix as

$$A.\beta_i = 0$$

Supply-demand constraint can be written as

$$\sum_{r=1}^{r_i} \beta_{ir} \geq d_i \qquad (II)$$

where d_i is the demand of part type i; and the non negativity constraint

$$\beta_{ir} \geq 0 \qquad (III)$$

Thus, the problem can now be expressed as the following min-time flow problem:

$$\text{minimize} \quad \sum_{u\epsilon U} \beta_i^u \tau_u^i$$

$$\text{subject to:}$$

$$(I), (II), \text{ and } (III)$$

The solution to this network flow problem provides us with values of β_{ir} for each part type.

Flexible Manufacturing Systems: Recent Developments
A. Raouf and M. Ben-Daya (Editors)
1995 Elsevier Science B.V.

Heuristics For Loading Flexible Manufacturing Systems

E. Kathryn Stecke and F. Brian Talbot

Graduate School of Business Administration,
The University of Michigan, Ann Arbor, Michigan, U.S.A.

Abstract

The flexible manufacturing system (FMS) is an alternative to conventional discrete manufacturing processes that permits highly automated, efficient, and simultaneous machining of a variety of part types. In managing these systems, technological requirements indicate that several decisions must be made prior to system start-up. To this end, previous research has defined a set of FMS production planning problems. The final production planning problem is called the loading problem, which is described as follows. Given a set of part types chosen for immediate simultaneous production, allocate the operations and associated tooling of these part types among the machines subject to the capacity and technological constraints, and according to some loading objective. This problem has previously been formulated as a nonlinear mixed integer program for several loading objectives. Although it has been shown that the nonlinear MIPs are solvable on large computer systems, real-time FMS control requirements and the typical availability of minicomputers in shop environments make it impractical and cost inefficient to optimally solve the loading problem in many plants today.

As a result, the authors develop several heuristic algorithms that provide good solutions to various versions of the FMS loading problem. We expect that these rules can be executed in essentially real-time on minicomputers available today.

1   Introduction

The development of highly automated flexible manufacturing systems (FMSs) has created new opportunities for the efficient manufacture of component parts in the metal-cutting industry. The effective use of these systems, however, requires the solution of new and complicated production planning and control problems. In an effort to make these problems tractable, Stecke [1983] has devised a hierarchical scheme comprising five production planning problems which must be solved prior to system operation. A brief description of these appears in Table 1. The primary purpose of this paper is to present heuristic solution procedures for one of these problems, the FMS loading problem.

The loading problem is one of deciding how individual machines are to be tooled to collectively accomplish all manufacturing operations for each part type that will be machined concurrently. A solution to this problem specifies the cutting tools which must be loaded in each machine's tool magazine before production begins, and hence, the machine or machines to which a part can be routed for each of its operations. Since a variety of products (parts) can be manufactured simultaneously on an FMS, where each part has its own, potentially unique, set of required operations, loading becomes a combinatorial problem. Some of the characteristics which make this a difficult combinatorial problem to solve include the possibilities that:

1. some particular part operations may be performed on any of several different types of machines;

2. operations could then require different processing times on various machine tools;

3. different part operations may be able to use some of the same cutting tools; and

4. tools, measured individually and collectively, require various amounts of space (slots) in fixed-size, limited-capacity tool magazines.

Table 1
Production Planning Problems

1. Part Type Selection: From a set of part types that have production requirements, determine a subset for immediate and simultaneous processing.
2. Machine Grouping: Partition the machines into machine groups in such a way that each machine in a particular group is able to perform the same set of operations.
3. Production Ratio: Determine the relative ratios at which the part types selected in problem (1) will be produced.
4. Resource Allocation: Allocate the limited number of pallets and fixtures of each fixture type among the selected part types.
5. Loading: Allocate the operations and required cutting tools of the selected part types among the machine groups subject to technological and capacity constraints of the FMS

Although it is possible to model these characteristics as nonlinear mixed integer programs (Stecke [1983]), it can be time and cost prohibitive to optimally solve the resultant loading problems that are large, despite the existence of an efficient branch and bound procedure that can solve the nonlinear integer problems associated with one of the 5 FMS loading objectives of interest here (Berrada and Stecke [1983]). Hence, there will be a need for fast heuristic procedures that give good, if not optimal, solutions for large scale FMS loading problems.

The loading problems addressed in the first section assume that the grouping problem ((2) of Table 1) has been solved. Stecke [1981] introduced the notion of grouping machines as one way of simplifying overall production planning and control of FMSs. Grouping also automatically provides machine redundancy, which is very useful during machine breakdown situations. A machine group is composed of machines that are tooled identically so that they can individually perform the same operations. Typically, machines in a group are of the same type and are identically tooled. Through the use of closed queuing networks to model FMSs, Stecke and Solberg [1982] found optimal grouping patterns and associated optimal allocation ratios, which indicate the amount of work (operation processing time) which should ideally be assigned to each machine group, to provide maximum expected production. Knowing how machines should be grouped affects other aspects of production planning, but specifically, simplifies the loading problem both by reducing the tooling options and by reducing the size of the problem to be solved. In Section 2, FMS loading heuristics are suggested that also group the machine tools.

Table 2
Alternative Loading Objectives

1.	Minimize part movement between machines, or equivalently, maximize the number of consecutive operations for a part to be processed by the same machine;
2.	Balance the workload (total processing time) per machine on all machines;
3.	Balance the workload per machine for a system configured of groups of machines of equal sizes;
4.	Unbalance the workload per machine for a system of groups composed of unequal numbers of machines;
5.	Duplicate certain operation assignment

Several loading objectives are considered in this paper and are indicated in Table 2. Each might be applicable in different manufacturing situations. In some systems, several objectives may apply.

2   Loading Heurstics with Grouping

In this section, several loading algorithms are described for the situations where the grouping problem has already been solved. That is, we know how many groups there are, the sizes of each machine group, and which machines are in each group. The solution to the loading problem will define precisely which operations, and hence tooling, will be assigned to each machine group. There are several machines of each type. (If there is only one machine of each type, then the loading problem that is solved in this section becomes trivial). We initially assume that each operation can be accomplished by only one machine type. This assumption can easily be relaxed.

2.1   FMS Loading Algorithms for Minimizing Part Movement

The first two algorithms are designed to minimize part movement through an FMS. This objective is especially important in a system having relatively high travel or pallet positioning times (see Stecke and Solberg [1981]) or if the material handling system is a bottleneck. The first algorithm, approaches the problem by examining consecutive operations sequentially, whereas the second, pre-groups consecutive operations before loading. Throughout, the algorithms are presented in increasing order of complexity.

Algorithm 1

1. Taking each part type in numerical order, assign each operation consecutively to the lowest numbered machine tool of the correct type which has magazine capacity available for the tools required for the operation.

2. Continue assigning operations until all have been allocated.

This is a simple application of the first-fit bin packing heuristic (see Johnson [1973]) and involves very little computational effort beyond feasibility testing. At each tool magazine capacity test, common cutting tool and tool slot overlap checks as well as corresponding adjustments, should be made to determine feasibility. A potential drawback of this simple, naive approach is that the resulting solution will likely not conform to given commonly-tooled machine grouping goals, related to total assigned processing times.

Algorithm 2

1. For each part type, group maximally into "operation sets", consecutive operations which require the same machine type. Calculate the number of magazine slots required for each operation set.

2. Calculate a priority index for each operation set, and assign operation sets to machines of the correct type according to this index. Several possible prioritizing schemes are:

(a) assign operation sets to the slowest numbered machine possible according to the index: "largest number of tool slots required" first. This is a variation of the first-fit-decreasing bin packing heuristic.

(b) assign operation sets according to the index "largest number of tool slots required" first, but to the machine having the cutting tools already in its magazine which will most reduce the number of slots actually needed by the operation set being assigned.

(c) assign operation sets to the lowest numbered machine possible according to the index: "largest number of operations in a set" first.

(d) assign operation sets according to the largest value of the ratio: (number of operations in a set)/(number of additional tool slots required). This rule is designed to assign as many operations as possible at the lowest cost in terms of additional tool magazine slots needed.

These heuristics of Algorithms 2 will, like Algorithm 1, probably give solutions which do not conform to ideal groupings of machines as provided by closed queuing network analysis. In addition, if the use of maximally grouped sets of operations does not lead to a feasible solution, then alternative methods must be devised to define operation sets. This could be a difficult problem, although as Stecke [1981] has suggested, a starting point for defining sets can be provided by the L.P. relaxed solution to the nonlinear I.P. statement of the FMS loading problem. The L.P. solution could be adjusted heuristically, to conform to the integrality requirement while remaining feasible.

2.2 Loading Procedures for Balancing and Unbalancing Objectives

Closed queuing network analysis provides idealized groupings of identically tooled machines as well as corresponding optimal group workload allocation ratios (Stecke and Solberg [1982]). In general, the analysis proves that for balanced configurations of grouped machines, expected production is maximized by balancing the assigned workload per machine. Alternatively, better machine configurations are unbalanced, and in these situations, expected production is maximized by a specific unbalanced allocation of work.

The following heuristics are designed to assign operations to machines in an effort to meet these optimal allocation ratios. These ratios were developed to provide guidelines on how to allocate work to machines to maximize expected system productivity as measured by the amount of processing time which can be completed in a given period of time.

Before the following loading algorithms are implemented, it is useful to obtain an estimate of the maximum workload per machine. This is accomplished with the following calculations:

1. Sum the operation processing times, weighted by the part production ratios, of all operations for all part types (that will be simultaneously produced by the

FMS over the next time period) that require a particular type of machine. Do this for each type of machine. (Part production ratios can be either provided by a production plan, or can be analytically derived ratios designed to maximize system productivity.)

2. Divide each of these sums by the corresponding number of machines of each type to obtain an estimate of the workload per machine, if the workload were balanced.

Algorithm 3

1. Order machine groups within each machine type by nonincreasing number of machines in each group (as previously solved by the grouping problem).

2. Order operations, for all part types that shall be simultaneously machined, which require the same machine type, by nonincreasing processing time.

3. Assign the first operation from each of the lists developed in Step 2, to the first machine group of the correct type.

4. The assignments of the remaining operations depend on whether the machine tools of a given type are organized into groups of equal or unequal size:

    (a) *Equal Size Machine Groups* (requires the Balancing Loading Objective).
    It is necessary to assign operations from their ordered lists to their corresponding required machine types. Thus, the following allocation procedure is repeated for each list. Suppose there are L machine groups for some list. Consecutively assign the first L operations from this list to the L machine groups. Consecutively assign the next L operations, but in reverse machine group order. For example, machine group one will contain operations 1, 2L + 1, 4L, 4L + 1, etc. Machine group two will contain operations, 2, 2L - 1, 2L +1, etc.

    (b) *Unequal Size Machine Groups* (required the Unbalancing Loading Objective)
    The following procedure is repeated for each operation list as they were defined in step 3: Corresponding to each list is a set of machine groups which were ordered in step 2. Suppose these groups are labelled: $\ell = 1, ..., L$, with the number of machines in the ℓth group denoted by M_ℓ. The allocation rule is to assign the next M_ℓ operations to the ℓth group, for $\ell = 1, ..., L$. When $\sum M_\ell$ operations have been assigned, continue the process with $\ell = 1, 2^\ell, ..., L$. Repeat until all operations have been assigned.

Figure 1 further illustrates the application of Algorithm 3(a). Here, nine machines of the same type have been placed into three equal sized groups. Operations one to nine have been allocated thus far, as indicated by the numbers in the rectangles. The height of each rectangle is proportional to the operation processing time. The rationale for

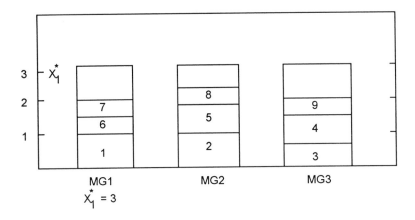

Figure 1: Three-Sized Groups Containing Nine Machines

reversing the allocation across groups, in an effort to balance workload, is evident from Figure 1. If the processing times are highly variable, then it may be worthwhile to deviate from the rigid allocation procedure by skipping a group that appears to have an excess workload, or allocating an extra operation to an underloaded group, or by making adjustments (i.e. pairwise exchanges) after an initial solution has been found.

Figure 2 illustrates Algorithm 3(b). There are seven machines of the same type placed into three groups of four, two, and one machine in each group. Operations one to 14 have been assigned thus far, as indicated by the numbered rectangles. The height of each rectangle is a measure of the operation processing time. The X_i^* values are hypothetical optimal allocation ratios which would usually be obtained from using the closed queuing network model. These ratios are guidelines which could be used to measure the "goodness" of a particular heuristic (as well as an optimal) solution. Consistent with the relative magnitude of these ratios, the heuristic procedure described aims to assign slightly more than an average amount of processing time to the larger groups and slightly less than average to the smaller groups.

A more comprehensive example demonstrating Algorithm 3 will now be given. Figure 3 illustrates a 13-machine manufacturing system containing three types of machines which have been arranged into six groups. Optimal allocation ratios X_i^* are specified for each group. For each type of machine, the groups have been ordered such that the largest groups are first, according to Step 1.

Table 3 contains the ordered operations for all parts and the type of machine required. As specified by Step 2, operations have been ordered by largest processing time first.

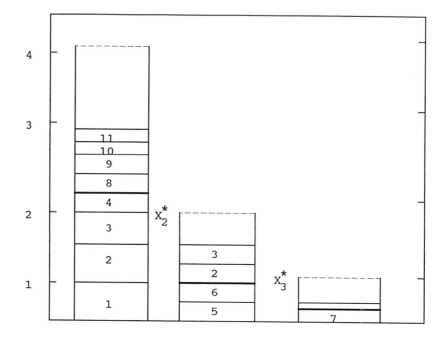

Figure 2: Seven Machines Partitioned Into Unequal Sized Groups

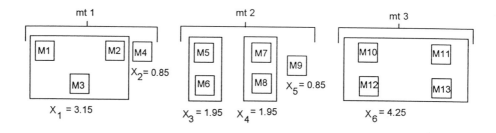

Figure 3: Thirteen Machines of Three Machine Types

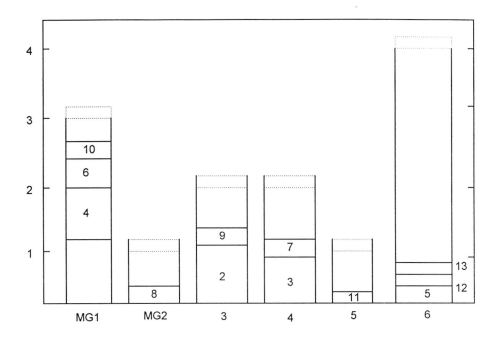

Figure 4: Allocation of Operations for the Thirteen Machine FMS

Table 3. Operations Ordered According to LPT First

Operations	1	2	3	4	5	6	7	8	9	10	11	12	13	...
Machine Type	1	2	2	1	3	1	2	1	2	1	2	3	3	...

Since the groups for machine type 1 are of unequal size, Algorithm 3(b) is followed. The number of operations assigned to each group is equal to the number of machines in each group. Then the process is repeated until all operations that require machine type 1 have been assigned. See Figure 4.

There are three groups of type 2 machines. Since groups three and four have equal numbers of machines and group five has a different number, a combination of Algorithm 3, Parts (a) and (b), is applied. Groups three and four will be assigned operations on the forward and reverse pass. Group five will be assigned an operation at the <u>end</u> of the reverse pass, as displayed in Figure 4. Machine type three has only one group of four machines, so all operations are simply assigned to it in order.

2 Loading Heuristics with Grouping

In this section, several loading algorithms are described for the situations where the grouping problem has already been solved. That is, we know how many groups there are, the sizes of each machine group, and which machines are in each group. The solution to the loading problem will define precisely which operations, and hence tooling, will be assigned to each machine group. There are several machines of each type. (If there is only one machine of each type, then the loading problem that is solved in this section becomes trivial). We initially assume that each operation can be accomplished by only one machine type. This assumption can easily be relaxed.

2.1 FMS Loading Algorithms for Minimizing Part Movement

The first two algorithms are designed to minimize part movement through an FMS. This objective is especially important in a system having relatively high travel or pallet positioning times (see Stecke and Solberg [1981]) or if the material handling system is a bottleneck. The first algorithm, approaches the problem by examining consecutive operations sequentially, whereas the second, pre-groups consecutive operations before loading. Throughout, the algorithms are presented in increasing order of complexity.

Algorithm 1

1. Taking each part type in numerical order, assign each operation consecutively to the lowest numbered machine tool of the correct type which has magazine capacity available for the tools required for the operation.

2. Continue assigning operations until all have been allocated.

This is a simple application of the first-fit bin packing heuristic (see Johnson [1973]) and involves very little computational effort beyond feasibility testing. At each tool magazine capacity test, common cutting tool and tool slot overlap checks as well as corresponding adjustments, should be made to determine feasibility. A potential drawback of this simple, naive approach is that the resulting solution will likely not conform to given commonly-tooled machine grouping goals, related to total assigned processing times.

Algorithm 2

1. For each part type, group maximally into "operation sets", consecutive operations which require the same machine type. Calculate the number of magazine slots required for each operation set.

2. Calculate a priority index for each operation set, and assign operation sets to machines of the correct type according to this index. Several possible prioritizing schemes are:

(b) If the job is either machine time or tool infeasible, the job is considered for splitting, either by splitting the batch size, or by splitting the operation sequence. Either type of splitting will likely result in process inventory, and a loss of batch continuity. Reducing the batch size, of course, has no effect on tool infeasibility, whereas splitting operations can reduce both machine time and tool infeasibility. The desirability of splitting would involve these trade-offs as well as the potential attendant loss of machine utilization if no splitting were permitted.

6. Revise priority indices for jobs remaining on the arrival list to reflect the current status of the system, and to include any new job which may have been added to the list. Repeat steps 1-6 until no jobs remain on the waiting list, or no additional jobs, or portions of jobs (from splitting), can be loaded.

In step 5, partial machine grouping may be accomplished or avoided depending upon what is desired. Grouping requires duplicate tooling, and hence, may be uneconomical. However, grouping reduces the disruption machine or tool breakdowns may have on the completion of a particular job, it effectively increases the production rate of the job, and it may be required to avoid job splitting. To illustrate how partial grouping can be accomplished, consider a job with a batch size of N parts. Suppose that each part requires four operations on machine type M, and that two machines of type M are available. The two machines could be tooled identically to perform all four of these operations or some subset of the four. Alternatively, if time and tool capacity exists, grouping could be avoided by providing one set of tools to one of the machines.

Algorithm 4 is a sequential procedure as stated, but it could be modified to more rigorously examine the interactions between jobs. For example, once jobs have been found machine time and tool feasible, they could be placed on an intermediate list. From this intermediate list, jobs could be further evaluated two at a time (or more) to take advantage of common tooling requirements, or batch splitting.

4   Summary and future research

This paper presents several heuristics for determining how machine tool magazines in a flexible manufacturing system can be loaded to meet simultaneous production requirements of a number of different part types. This loading problem is multicriteria in nature, and hence, no one of the heuristics introduced would like meet the needs of all FMSs. Future research is needed to better define the variety and character of FMS loading objectives, how the loading problem links with the other four FMS production planning problems presented, and how loading and real time scheduling of parts on a system interact. It seems that detailed simulation of real systems could be used to help analyze these issues.

References

[1] Mohammed Berrada and Kathryn E. Stecke, "A Branch and Bound Approach for Machine Loading in Flexible Manufacturing Systems," Working Paper No. 329, Division of Research, Graduate School of Business Administration, The University of Michigan, Ann Arbor MI (April 1983).

[2] D.S. Johnson, "Near Optimal Bin Packing Algorithm," Ph.D. dissertation, Mathematics Department, M.I.T., Cambridge MA (1973).

[3] Kathryn E. Stecke, "Experimental Investigation of a Computerized Manufacturing System," Master's Thesis, School of Industrial Engineering, Purdue University, W. Lafayette IN (December 1977).

[4] Kathryn E. Stecke, "Production Planning Problems for Flexible Manufacturing Systems," Ph.D. disertation Department of Industrial Engineering, Purdue University, W. Lafayette IN (August 1981).

[5] Kathryn E. Stecke, "A Hierarchical Approach to Production Planning in Flexible Manufacturing Systems," in *Proceedings, Twentieth Annual Allerton Conference on Communication, Control and Computing*, Monticello IL (October 6-8, 1982).

[6] Kathryn E. Stecke, "Formulation and Solution of Nonlinear Integer Production Planning Problems for Flexible Manufacturing Systems," *Management Science*, Vol. 29, No. 3, pp. 273-288 (March 1983).

[7] Kathryn E. Stecke and Thomas L. Morin, "Optimality of Balancing Workloads in Flexible Manufacturing Systems", *International Journal of Operational Research* (1984), forthcoming.

[8] Kathryn E. Stecke and James J. Solberg, "Loading and Control Policies for a Flexible Manufacturing System," *European Journal of Production Research*, Vol. 19, No. 5, pp. 481-490 (September-October 1981).

[9] Kathyrn E. Stecke and James J. Solberg, "The Optimality of Unbalancved Workloads and Machine Group Sizes for Flexible Manufacturing Systems", Working Paper No. 290, Division of Research, Graduate School of Business Administration, The University of Michigan, Ann Arbor MI (January 1982).

Flexible Manufacturing Systems: Recent Developments
A. Raouf and M. Ben-Daya (Editors)
© 1995 Elsevier Science B.V. All rights reserved.

A Framework For Developing Maintenance Policy For Flexible Manufacturing Systems

A. Raouf
Systems Engineering Department, King Fahd University of Petroleum and Minerals
Dhahran 31261, Saudi Arabia

Abstract

Unique characteristics of FMS from a maintenance management point of view are presented. Pertinent literature is briefly outlined. A step-by-step methodology for developing a maintenance policy for FMS is presented.

1 Introduction

Maintenance activities are carried out to maintain a system in as-built condition so that it keeps its original production capacity and quality capability. Increased competition is turning manufacturers to adopt automation. In the early days automation resulted in increased productivity at the cost of variety of products. To meet customer demands, i.e. reduced lead times, a new generation of manufacturing systems came into existence which are both highly productive and flexible. Such systems have come to be known as Flexible Manufacturing Systems (FMS).

A flexible manufacturing systems can be defined as a collection of manufacturing units interconnected with material handling equipment under the supervision of one or more executive computers. Such a computer is usually programmed to control the machine operation, the scheduling of parts through the machining system, the accumulation of important data including the tool wear with each part program and failure diagnosis. Flexibility of a manufacturing system is the ability to respond to changes in product, product mix, process and environment. A brief outline of these factors is shown in Figure 1.

A typical FMS is a combination of complex machinery interfaced with an equally complex computer system. Since flexibility is a 'trend' and not a state the complexity of machines and the complexity of computer systems are ever increasing. This results in an increased capital outlay. Each machine itself is a combination of many parts where these parts are inter-dependent. To be able to obtain maximum benefits out of FMS the management is attaching high importance to the maintenance management of these systems. Traditonally maintenance dealt with "stand alone" machines or machines coupled together having similar mechanical parts. A large volume of literature covering

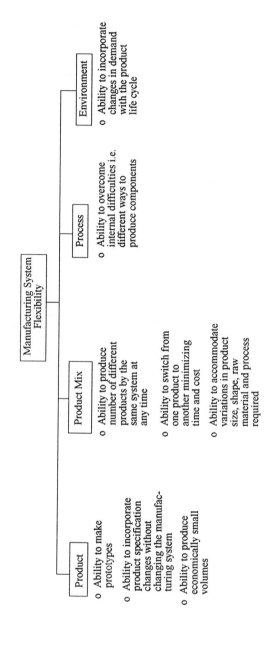

Figure 1: Manufacturing system flexibility factors.

various aspects of traditional maintenance is available. FMS maintenance management requires a different approach than traditional maintenance. This aspect does not seem to have been adequately addressed.

This chapter describes some of the unique aspects of FMS which suggest that renewed attention must be given to the maintenance of such systems. In addition, a very brief literature survey covering traditional maintenance in figure form is provided. Some of the recent work in respect of FMS maintenance are presented as well. A step by step methodology is presented which can be helpful in establishing maintenance policy for FMS.

2   Unique Aspects of FMS

Following are some of the unique aspects of FMS which suggest that renewed importance be given to their maintenance management.

2.1   Single component failure

Some of the salient features of FMS are its ability to produce a number of items and its ability to overcome internal difficulties caused by unforseen and unpredictable disturbances such as machine down time problems etc. These characteristics result in a minimum level of work-in-process (WIP) inventories. In cases of progressive manufacture, if the machine at the beginning of the process has a failure of one component, it may not cause the loss of output from this machine only, but due to a reduced WIP inventory the entire production process can quickly become idle.

2.2   Life Expectancy

FMS is able to respond to changes in product, product mix, process and environment and thus assist in achieving increased productivity and profitability for mid-volume, mid variety manufactures. The price of increased flexibility is increased complexity of machines and control systems. Since an FMS is programmable its life expectancy is greater than dedicated manufacturing systems. For example, a dedicated manufacturing system for manufacturing automobiles need new design modifications etc. every year so that it can produce the new models. This is not the case with an FMS. Management uses long life expectancy as a justification for high costs of FMS using life cycle cost approach.

2.3   System Utilization

To have a favourable return on investment (ROI) utilization of FMS is higher than conventional manufacturing systems. A utilization greater than 80% is not an uncommon feature of FMS. This necessitates of high levels of maintenance effectiveness.

2.4   Synergistic Costs

FMS is designed to provide a faster response to customers requests and reduce production control related losses. In fact the cost of an isolated failure not only reflect the loss of that piece of equipment but resulting synergistic costs like customer dissatisfaction etc.

From the above it can be seen that maintenance policies capable of keeping FMS in as-built condition are required. These policies must be capable of handling complex systems comprising of many dissimilar components. A typical FMS in addition to having mechanical parts also has electronic, hydraulic, electro-mechanical software and hardware elements and each has different failure characteristics and different levels of interdependencies.

3   Literature Survey

Articles related to maintenance have appeared in different journals over the past years. A brief citation of the work available in the literature is given in this section.

3.1   Conventional production system maintenance

A considerable volume of literature is available dealing with traditional manufacturing systems. Since providing a comprehensive review of the literature is not needed, important work is cited in Figure 2.

3.2   Automated Production Systems

Published work on maintenance policies in highly integrated, computerized manufacturing environment is scarce (Lie et al, 1977), (Sherif and Smith 1981).

Kennedy (1987) has raised several issues in planning FMS maintenance. The issues identified are:

1. self-diagnostic equipment/requirements and justification.

2. amount of preventive maintenance to be performed.

3. calculating down time costs.

4. in house maintenance capacity and maintenance to be carried out by outside contractors.

5. stock levels of spares.

He suggested that models be developed for minimizing total costs but did not develop actual model(s).

Vineyard (1990) tested five traditional maintenance policies in typical FMS i.e. corrective 30 days, preventive 90 days, opportunistic on failure, and opportunistic on

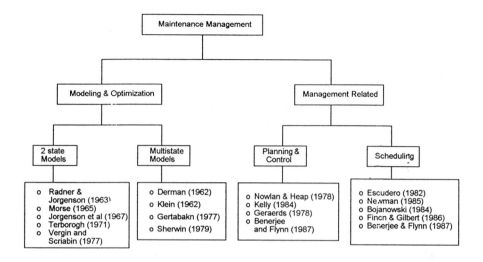

Figure 2: Literature citations for conventional system maintenance.

failure after 30 days. Using flow time, capacity, machine down time, number of maintenance tasks required and equipment utilization as criteria and using a discrete event network simulation model representing an existing FMS, he observed that there was a statistically significant difference in the maintenance policies with regard to all the performance measures. He further concluded that choice of a particular maintenance policy would affect an FMS system.

4 Step-by-Step Methodology

The literature survey reveals the lack of adequate methods for addressing FMS maintenance problems. In this section, we provide a step by step methodology for developing FMS maintenance policies. The objective of this methodology is to reduce the number of breakdowns by developing a preventive maintenance program. It is assumed that record of failures in respect of the system or for a similar system are available.

4.1 Methodology Outline

The suggested methodology is described by a flow diagram given in Figure 3.

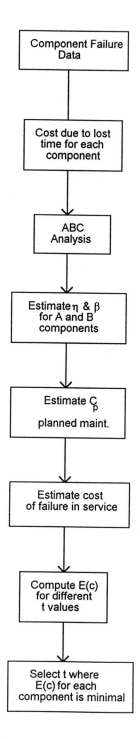

Figure 3: Step-by-step methodology.

4.1.1   Failure Data Record

The data used for developing the maintenance plan is a record of failures of the existing
FMS components or similar system operating at some other location. The data usually
gives the following details

- the date of each event

- the length of time out of service

- the nature of the breakdown and the description of repair work done

- part and equipment identification of the failed component.

4.1.2   ABC Analysis Of Failure Data

To provide maintenance an efficient and a high return on investment on these services,
ABC or "Pareto" analysis is used. This analysis is based on a classification of failures
in terms of costs, usually hours or minutes of down time. This is used to give an order
of priority among the maintenance actions to be taken.

   The method starts by listing the machines in order of decreasing costs in down
time with the number of failures of each machine and forming the cumulative sums
of the costs and the corresponding failures. The cumulative cost is plotted against
cumulative failures, both expressed as percentage of the respective totals. Zone A
of this plot consists of nearly 20% of the failures that account for 80% of the costs.
Failures falling in this category are given top priority for maintenance and repair. In
zone B are contained the 30% of the failures that account for nearly 15% of the costs.
The remaining 50% of the failures that account for 5% of the costs are contained in
Zone C. This analysis identifies the failures that cause the most disruption to the
system.

   As mentioned earlier components of an FMS have different levels of interdependencies and as such care should be exercised in estimating the cost of time lost for each
component.

4.1.3   Estimation of failure distribution parameters

Vineyard (1990) has shown that with the exception of electrical failures, which had
a lognormal distribution, all other failures namely hydraulic, mechanical, electronic,
human error, and software related failures have weibull distribution.

   For the sake of simplicity and generalization we may assume that all failures have
a weibull distribution.

   A typical weibull probability density function is shown in Figure 4.

   The density function of weibull distribution is

$$f(t) = \frac{\beta}{\eta} \left(\frac{t}{\eta}\right)^{\beta-1} exp\left[-\left(\frac{t}{\eta}\right)^{\beta}\right]$$

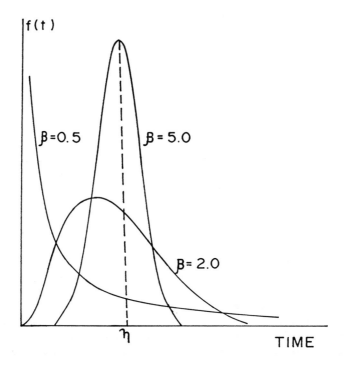

Figure 4: Weibull Probability Density Function

for $t \geq 0$ and reliability at time t is

$$R(t) = 1 - F(t) = exp\left[-\left\{\frac{t}{\eta}\right\}^{\beta}\right]$$

and failure rate is

$$\lambda(t) = f(t)/R(t) = \left(\frac{\beta}{\eta}\right)\left(\frac{t}{\eta}\right)^{\beta-1}$$

For each component contained in zone A, calculate estimates of $\eta \& \beta$. A computer package can be easily used to obtain these estimates. These estimates can also be obtained graphically and this method is explained in details in Jardine (1981). The estimate of $\eta \& \beta$ for every component can be estimated and summarized in a table form. If $\beta < 1$ then $\lambda(t)$ is a decreasing function of t. For $\beta = 1$, $\lambda(t)$ is constant and is equal to $\frac{1}{\eta}$ and if $\beta > 1$, $\lambda(t)$ is an increasing function of t. This phenomenon is shown in Figure 5.

Life Cycle of an equipment

In Figure 5 the failure rate $\lambda(t)$ is falling as t increases. This is the area where $\beta < 1$. This can be the result of malfunctioning in manufacturing processes. $\lambda(t)$ becomes nearly constant where the machines have been properly run-in. Part (b) of the figure may be considered as the maturity period where $\lambda(t)$ is constant. This usually is the case for electronic systems. Although, for mechanical system $\lambda(t)$ increases

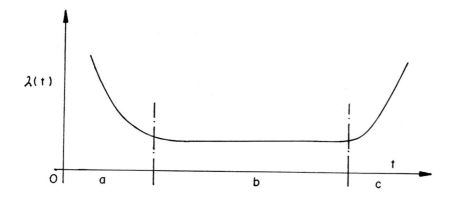

Figure 5: Life cycle of physical equipment (bathtub curve).

slightly as t increases, it is considered to be constant unless this increase is statistically significant. During part (c), $\lambda(t)$ is increasing and this makes monitoring of the equipment necessary and condition based maintenance may have to be resorted to.

Based upon these estimates $R(t)$ for each component can be computed. Suppose β was found to be 3 and $\eta = 600$ than

$$R(t) = exp\left[-\left(\frac{t}{600}\right)^3\right]$$

Then the cost of maintaining the system is made up of a fixed cost C_p and a variable C_r where $C_r = nC_f.f(n)$, $f(n)$ is the probability of n failures occuring during time periods 0 to T.

4.1.4 Cost estimates

For an on-going system the cost of planned replacement of a component is known. The cost of replacement of the component if it fails, which is considerably higher than the previous one, is known as well. Knowing the intervals at which preventive replacements should be performed yielding minimum cost can be determined.

C_p = cost of planned replacement.
C_p = cost of failure and replacement
$C_f = C_p$

4.1.5 Expected Cost

Let us assume that we decided to replace an item every T units of time. Then cost of maintaining the system is a fixed cost C_p (replace at the end of the period) and a

192

variable cost $nC_f.f(n)$ where $f(n)$ is the probability of n failures occurring during time period $0 - T$. Considering that probability of one failure during period T is greater than probability of two failure during period T thus probability of at least 1 failure = $1 - R(t)$ (Layonnet, 1991).

Let $E(c)$ be the expected cost of replacing a component during the period $0 - T$, then

$$
\begin{aligned}
E(c) &= C_p + \{1 - R(T)\}\, C_f \\
&= C_p + \left\{1 - exp\left\{-\left(\frac{\beta}{\eta}\right)\right\}\right\} C_f
\end{aligned}
$$

the estimates of β & η for components have already been obtained in 4.1.3.

4.1.6 Minimizing $E(c)$

$$
E(c) = C_p + \left\{1 - exp\left\{-\left(\frac{t}{\eta}\right)^{\beta}\right\} C_f\right\}
$$

In this function C_p, β, η & C_f are known. Varying a value of t pertinent $E(c)$ can be easily computed. For a starter mean time between failures (MTBF) of each component can be used. By varying t above and below MTBF t can be determined that minimize $E(c)$. A reasonable policy of preventive maintenance for each component can thus be set up.

5 Conclusion

An interactive, personal computer based package, based on the above methodology is undergoing development and testing stages. It is anticipated that maintenance managers will have an operational methodology which nearly minimizes the $E(c)$. The assumptions regarding weibull distribution having location parameter equal to zero and also that failures follow weibull distributions can easily be justified.

References

[1] Banerjee, A. and Flynn, B.B. (1987), "A simulation study of some maintenance policies in a group technology shop", *International Journal of Production Research*, **25** (11) pp. 1595-1609.

[2] Bojanowsk, R.S. (1984), "Improving Factory Performance with Service Requirements Planning (SRP)", *Production and Inventory Management*, **25** pp. 31-44.

[3] Derman, D., (1962) "On Sequential Decisions and Markov Chains", *Management Science*, **9**, pp. 16-24.

[4] Escudero, L.F., (1982), "On maintenance scheduling of production units", *European Journal of Operational Research*, **9**, pp. 264-274.

[5] Finch, B.J., and Gicbert, J.P. (1986), "Developing Maintenance Craft Labor Efficiency Through An Integrated Planning and Control System : A Perspective Model", *Journal of Operations Management*, **6**, pp. 449-459.

[6] Geraerds, W.M.J., (1978), "Trends in Maintenance Strategies and Organization : Maintenance Philosophies - State of the art", *4th EFNMS Congress*, London, 4-3 (i) - (ii).

[7] Gertabakh, I.B. (1977), *Models of Preventive Maintenance*, North Holland, Amsterdam.

[8] Kelley, A., (1984) *Maintenance Planning and Control*, London.

[9] Klein, M., "Inspection - Maintenance - Replacement Schedules under Markovian Deterioration", *Management Science* **9** , pp. 25-32.

[10] Jardine, A.K.S. (1981) *Maintenance, Replacement and Reliability*, Pitman, London.

[11] Jorgenson, D.W. and McCall, J.J., (1967) "Optimal Scheduling of Replacement & Inspection", *Operations Research*, **11**, pp. 732-740.

[12] Lie C.H., Hwang, C.L. and Tillman, F.A., "Availability of Maintenance Systems : A state of the art survey", *AIIE Transactions*, **9** (7) pp. 247-259.

[13] Kennedy Jr. W.J. (1987), "Issues in the Maintenance of Flexible Manufacturing System", *Maintenance Management International*, **7**, pp. 43-52.

[14] Layonnet, P., (1991), *Maintenance Planning, Methods and Mathematics*, Chapman & Hall London.

[15] Morse, P.M. (1965), *Queues, Inventories and Maintenance*, Wiley, New York.

[16] New Man, R.G. (1985), "MRP where M = Preventive Maintenance", *Production and Inventory Management*, **2**, pp. 21-28.

[17] Nowlan, P.S. and Heap, H.F. (1978), "Reliability Central Maintenace", National Technical Information Service Report AD-A066-579.

[18] Radner, R. and Jor Genson, D.W., (1963), "Opportunistic Replacement of a Single Part in the Presence of Several Monitored Parts", *Management Science*, **10**, pp 70-84.

[19] Sherif, Y.S. and Smith, M.L., "Optimal Maintenance Models for Systems Subject to Failures : A Review", *Naval Research Logistics, Quarterly*, **28**, pp. 47-74.

[20] Sherwin, D.J., (1979), "Inspection Intervals for Condition Maintenance Items which Fail in an Obvious Manner", *IEEE Transactions Reliability*, **28**, pp. 85-89.

[21] Terborogh, G.A., "Practical Method of Investment Analysis", The MAPI System, Machinery and Allied Products Institute & Council for Technological Development, Washington, D.C.

[22] Vineyard, M.L., (1990), A Comparison of Maintenance Policies in Terms of Their Effectiveness in a Flexible Manufacturing Environment, Unpublished Ph.D. Dissertation, University of Cincinnati.

[23] Vergin, R.C. and Scriabin, M., (1977), "Maintenance Scheduling for Multi-component Equipment", *AIIE Transactions*, **9** (3) pp. 297-305.

Part IV

FMS Control

Flexible Manufacturing Systems: Recent Developments
A. Raouf and M. Ben-Daya (Editors)
© 1995 Elsevier Science B.V. All rights reserved.

FMS Planning and Control - Analysis and Design

Christian Bérard[a], Lucas Pun[b], Bernard Archimède[c]

[a]Maître de conférence au LAP/GRAI
Université Bordeaux I, 351 Cours de la libération 33405 Talence
(Tel : (33).56.84.65.30 - Fax : (33).56.84.66.44)
Email: berard@lap.u-bordeaux.fr

[b]Professeur honoraire au LAP/GRAI

[c]Maître de conférence à l'École Nationale d'Ingénieur 65000 Tarbes

Abstract :

In this paper FMS Planning and Control are regarded as Control problems. Control terminology is used and sometimes re-defined to analyse and design a Convivial FMS Planning and Control System. The objective is to have a unified methodology leading to a specific software. After defining the general concepts, the Analysis phase is concerned with the Modelling tools and techniques used to state the Hierarchical Planning System (called Predictive Control) and the Control System (divided into On-line Control and Dynamic Control). Then the Synthesis phase defines the method to specify and build the final software. This software is designed so that all decisions may be made through a dialogue between decision-maker and computer. We call it a Convivial Control.

1. Introduction and problem formulation

The aim of this work is to propose an Analysis and Design Methodology to control FMSs. That leads to specify a Software Architecture allowing to plan and control FMS operation.

According to the GRAI terminology [DOU.84], the analysis is based on the two fundamental concepts of System and Activity. An FMS is composed of two systems : the Physical System and the Monitoring System. Each of them is split into sub-systems and each sub-system into activities. Two sets of activities are considered : Decision Activities when a choice has to be made by a human operator, and Execution Activities when the operation may be described by a deterministic algorithm. In order to integrate FMS operation in the overall operation of the manufacturing system, FMS management is located on a Decision Level closely related to the upper level.

In order to structure our Analysis and Design Methodology we have split the Monitoring System into three parts : Predictive control, On-line control and Dynamic control. To formulate the reaction to disturbances we re-define a number of terms commonly used in control literature. In fact, controlling a machine is similar to controlling an FMS. But controlling an FMS makes it necessary to specify a more complex technological structure, and to take into account human decisional factors, various and numerous disturbances and multiple objectives to reach. This is why we choose to widen the scope of some definitions.

To detect and diagnose the origin of a disturbance, we propose to use Flow-Profiles presented in [ARC.91-1]. The consequences of a disturbance on the plan are calculated by means of the method of the Minimal Potential-Influence Zone also presented in [ARC.91-1].

The main aim of this paper is to solve FMS control problems convivially. This means that the control software is open to decision-makers : they can enter new constraints and obtain a view of the FMS true state through synthesised graphic diagrams. The co-operation between the decision-makers and the control software is obtained through these interactive diagrams working like input-output devices. Input is used by operators to transmit their choices and output gives information on the situation of the problem-in-process.

1.1. Role of an FMS in a Discrete Production System

An FMS is a set of machines and equipment which can be reorganised each time a new production program is launched.

The objective is twofold :

> - fast adaptation to multiple product types and multiple production program situations,
> - maximum rate of utilization (100% if possible) of the machines and the equipment.

To achieve this aim, two important problems must be solved :

> - the procurement of materials and articles must suit very closely the manufacturing needs,
> - the planning and control of the FMS activities, which become extremely complex, must be very efficient.

According to the objectives defined by the long-term management, the medium-term management takes care of two main tasks:

> - first, define and realize the procurement program,
> - second, define the manufacturing program which is realized by the FMS.

Therefore, the management of the FMS is a short-term planning and control task.

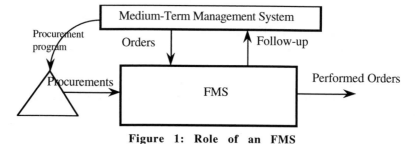

Figure 1: Role of an FMS

Each order o_i received by FMS has two dates:

> - d_i the earliest starting date,
> - f_i the latest ending date.

The main objective assigned to the FMS by the MTMS (Medium-Term Management System) is to perform all the orders and meet these two dates. The role of MTMS is to determine the triplets (o_i, d_i, f_i) through a hierarchical planning structure described in § 2.1.2.1.

1.2. Role of an FMS Monitoring System

The management of an FMS is performed by a **Monitoring System**. Its objective is to match :

> - on the one hand, the manufacturing programs required by the medium-term planners.
> - on the other hand, the realization of these manufacturing programs by the FMS.

Practical planning tasks :

> - detailed analysis of the production program coming from the medium-term according to the TPP (Technical Process Planning),
> - evaluation of the manufacturing load; if the manufacturing load is found insufficient or too important, FMS has to react and ask the MTMS for a modification
> - elaboration of a first schedule on the basis of the utilization of the main machines
> - analysis of availability of the logistics (pallets, fixtures, tools, conveyors, set-up problem,...)
> - readjustment of the first schedule according to the results of the analysis (using a Discrete Event Simulation tool) [BER 85], [ARC.87].

Scheduling tasks have to be flexible according to the amount of time available to obtain a good solution. Strategies and algorithms have to be chosen so that a quick solution may be found. If there is enough time the planning system tries to improve it. Improvement of scheduling solutions may be obtained through an algorithm or by a man-machine co-operation, often by both. We try to determine what strategy, what algorithms and what interfaces are convenient to perform such scheduling tasks.

Practical control tasks :

> - extraction of some operations from the production program according to the execution of the plan,
> - dispatching them,
> - elaboration of the "moving" strategy,
> - aquisition of "events" from the machines and equipment,
> - comparison of FMS true state with plan,
> - reaction and, if necessary, modification of the plan.

1.3. FMS Planning and Control viewed as Control problems.

In terms of control, FMS planning and control can be divided into three parts each having distinct tasks:

(a) **Predictive control** : elaborates adequate schedules to guide the FMS operations.

(b) **On-line control** : supervises the FMS operation and solves small problems caused by minor disturbances.

(c) **Dynamic control** : helps short-term planners to solve big problems caused by important disturbances.

Those three types of control are designed to be convivial. There is a big difference between automatic control and convivial control. In automatic control, the control actions are completely predetermined. In convivial control, the real-time control actions are elaborated by exploiting both the natural intelligence of the human monitor, and the artificial intelligence implemented in the computer system, so that jointly and convivially, the planning and control can be performed efficiently.

In such a view of FMS Planning and Control, it is possible to state a generic structure linking the three types of control defined above. Two criteria allow to define them globally:

- running mode: **periodic**, **event-driven** or both,
- decision level: short or very short term.

In this structure scheduling techniques are used in Predictive Control of course but they are also used in Dynamic Control to react and find a new schedule. The difference is that Predictive Control calculates a global new plan periodically, while Dynamic Control calculates a partial new plan each time a significant disturbance is detected. So, the plan is adjusted as many times as necessary. This plan is used to guide the FMS towards its objectives; it is called the **Reference Plan**.

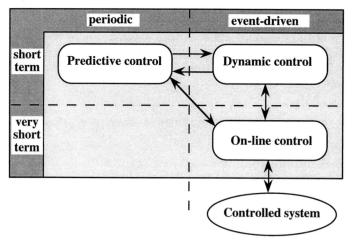

Figure 2 : FMS Planning and Control viewed as Control problems

To analyse those three control tasks it is important to define more precisely what a disturbance is and what the types of disturbances and the types of plan are ?

1.3.1.   Disturbances

A disturbance is an event. It is detected by a difference measured between two events : the first one is predictive and calculated in the scheduling task, the second one is acquired from the controlled system. The events may be evaluated from several points of view **occurrence date, quantity of product, type of product**, quality of a product. The problem of the quality of a product is equivalent to measuring a quantity of product obtained after a control operation on the product. A disturbance is a more general concept than the error concept (as usual in control literature); for us, an error is an evaluation of some disturbances.

To define a typology of disturbances , we use two criteria:

- the time required to react,

- the consequences on the **Reference Plan** if no decision is made. This plan obtained after a disturbance without adjustment is called the **Followed Plan**.

So, to define a typology of disturbances , we have to define the various types of plans.

1.3.2.   The various types of plans

In an ideal FMS operation, the **Reference Plan** is identical to the **Followed Plan**.

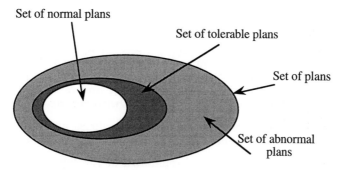

Set of normal plans

Set of tolerable plans

Set of plans

Set of abnormal plans

Figure 3 : The various types of plans

In a perturbed context, the **Followed Plan** is more or less far from the objectives defined in the **Reference Plan**. Three sets of plans are considered inside the whole set of possible plans:

- The normal plans meet the main manufacturing objectives (to perform all the orders and respect the two dates d_i and f_i). The **Reference Plan** is an instance of this set.

- **The tolerable plans** meet a set of tolerable objectives (to perform <u>a sub-set</u> of the orders and respect the two dates d_i and f_i with a margin m_i). All normal plans are tolerable.

- **The abnormal plans** are non tolerable plans.

1.3.3. Typology of disturbances

Four types of disturbances may be distinguished:

- The *incidents* need a light adjustment of the **Reference Plan**. But the **Followed Plan** is still a normal plan. The time required to react is not important. On-line control is concerned by incidents.

Example: the duration of an operation is increased because of a machine or another equipment breakdown but the difference is inferior to the predictive margin of the operation.

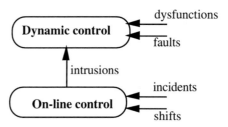

Figure 4 : Typology of disturbances

- The **intrusions** are easily detected but need a sophisticated and time-consuming algorithm to determine the consequences. The **Followed Plan** is abnormal. The *intrusions* are treated by Dynamic Control.

Example: machine breakdown whose duration is superior to the predictive margin of the operation.

- The **shifts** need not be treated immediately. Generally, the **Followed Plan** is still tolerable. An accumulation of several shifts may make it abnormal. In this case, the sum of the shifts is called a **fault** and must be treated by the Dynamic Control. On-line control is concerned with shifts .

Example: small differences between Reference dates and **Followed** dates. These differences may be compensated by shifting lightly the next operations.

- The **dysfunctions** are due to a breakdown of the follow-up system. Their treatment may take a long time. The **Followed Plan** is abnormal. The **dysfunctions** are treated by the Dynamic Control.

Example: a human operator has forgotten to report the end of an operation.

1.3.4. On-line control

To supervise FMS execution and solve small problems caused by minor disturbances, the On-line Control is concerned with two main activities:

- Dispatching orders and moving products and some equipment,

- Managing two types of disturbances: shifts and incidents. These disturbances are treated by means of time shifting and techniques using permutable operations group [ROU.88] or with a Discrete Event Simulation tool driven by a rule-based expert system [SUN.89].

1.3.5. Dynamic control

To help medium-term planners solve big problems caused by important disturbances, the Dynamic Control has to act as a supervisor for the On-line Control. It uses two tools presented in § 2.1.3 and 2.1.4 :

- Flow-Profiles,
- Manufacturing Environment Multi-graph.

The Flow-Profiles are used to detect and diagnose disturbances. The Manufacturing Environment Multi-graph is used to determine the Minimal Potential-Influence Zone of the plan to re-schedule in order to react efficiently. These tools are presented in detail in [ARC.91-1].

1.3.6. Predictive control

To be able to guide the FMS towards its objectives, the FMS Control must avoid the Control Reverse phenomenon. This means when the Controlled FMS is no longer guided towards its objectives by the FMS Control. In that case the FMS Control only records the running of the FMS without being able to guide this running. This phenomenon occurs when the FMS schedule is not feasible because of the lack of accuracy of the model or because numerous and important perturbations imply too many constraints. The hierarchical planning structure aims to avoid this Control Reverse phenomenon.

The main objective of Predictive Control is therefore to calculate a set of possible schedules. This activity tries to elaborate schedules, with the following constraints:

- meet the date and quantity constraints given by the MTMS,
- run inside a limited amount of time,
- propose to the MTMS to relax some constraints if there is a risk of Control Reverse phenomenon,
- accept interactive modifications made by a human operator,
- maximise flexibility of schedules, i.e. minimise their disturbances sensitivity,

1.4. Development of the Convivial Control.

The development of this convivial control requires analysis and synthesis (design of the computer systems). The analysis helps to understand, to state the control problems clearly and to solve them. The synthesis allows to build the various components of the computer system (hardware and software), and more particularly, to define an architecture and a methodology to program the software. One important objective of the software is to be convivial, therefore analysis must state the interactions between man and computer and synthesis has to elaborate a good man-machine interface.

2.   Analysis and Problem Solving

In the analysis phase, we propose modelling tools and the procedures to use them, in order to solve FMS planning-control problems.

First, we present the basic concepts to model the physical system or the FMS controlled system. This model falls into two parts : a kernel, used by all the functions of the Monitoring System, and added elements specific to planning functions.

Then, modelling tools for the control system (the monitor) are presented. As our work focused mainly on Predictive Control and Dynamic Control, we have split these tools into three parts : Predictive Control, Auscultation and Diagnosis for Dynamic Control and Therapy also for Dynamic Control.

Modelling tools are used through procedures. Those describe the way these tools may be applied to solve planning-control problems convivially.

2.1.   Modelling Languages and Techniques.

The first thing to be modelled is the FMS. For that purpose, we use typical concepts of Discrete Event Model and we structure them with an Object Oriented Approach. Three types of elements are considered : Product, Resource and Time. According to the control function using these concepts, the detail level is not the same; so, we organise this model around a kernel.

Then, Predictive Control Model outlines the concepts and principles of the Hierarchical planning and discusses the coordination strategy between the Decision Levels.

To inspect the FMS operation, we use Flow-Profiles as a modelling tool to reduce the number of points to be observed in order to detect disturbances. Flow-Profiles are also efficient to locate the causes of these problems.

Manufacturing Environment Multi-graph allows to analyse the effect produced by a disturbance on the reference plan. It determines the Minimal Potential-Influence Zone (MPIZ) of the plan to re-schedule.

2.1.1.   Modelling the controlled system

Three types of elements have to be modelled : Product, Resource and Time.

Product is a generic name for all the material processed by the FMS. It concerns the modelling of :

> - the description of the flows of products concerned by the FMS, description of main flows, description of detailed interactions between products, main and secondary resources,
> - the TPP (Technical Process Planning),

Resource is a generic name for all the equipment used to process the products.

It concerns the modelling of :

- the main resources: machines,
- the secondary resources: (pallets, fixtures, tools, conveyors,...); for each of them, the description of their interactions with main resources and products,
- the information between physical elements allowing to specify dynamic behaviour.

Time is a generic name for the elements defining horizon, period, calendar, interval of time.

All these elements need:

- a description of the data defining their static and dynamic attributes,
- a description of the rules defining their dynamic behaviour according to the state of the other interrelated elements.

Two types of dynamic behaviour have to be specified separately:

- technological behaviour linked to the physical structure of the modelled element,
- managerial behaviour linked to some choices on the way to manage locally the element.

Stating these two dynamic behaviours separately is not a trivial task. So, we propose to use concepts commonly used in Discrete Event Model [MAD.90] and some used in Object Oriented Programming [GRA.91].

The modelling task has to emphasise a number of elements according to the Control phase. But modelling tools have to be coherent to assume a coherent Information System and therefore a good data base for the FMS.

Consequently, we propose to build **a common kernel** defining the common elements of the three phases of the Control [MER.93]. Then we define some added elements for each particular phase.

All elements of the model are objects. In many standard Class Libraries the class of this object is called a Collection. For each object, you have:

- an identifier,
- pointers to navigate in the list of same type objects,

2.1.1.1. Common kernel model

We will not give an exhaustive description of the kernel but only some examples to explain how to build it.

Regarding the modelling of the **Time concepts**, we have to define the calendars used in the FMS. Calendars allow to define precisely the nominal availability of each resource. A Calendar is built from the basic concepts : Day Of The Week, Hour, HourInterval, WeekCalendar, Days Of Rest and Date. All are objects and are defined with their own attributes and operators.

For instance,

> List of WeekCalendars. For each WeekCalendar, one has:
> - seven lists of HourInterval (one for each Day Of The Week), each HourInterval is a couple of Hours (HourD, HourF).
>
> List of Calendars. For each Calendar, one has:
> - a list of Days Of Rest,
> - a list of (WeekCalendars, DateD, DateF).

Regarding the modelling of the **Resource concepts**, we have to define several **classes of resources** (Machine, Operator, pallets, fixtures, tools, conveyors,...) and several **hierarchical structures** (Group, Partition, Layout). Hierarchical structures are useful to define dependence relations between resources. **Group relation** defines sets of equivalent resources. **Partition relation** is used to settle split resources (for instance, a work table with three places, some products need the whole table, others only one place). **Layout relation** describes an inclusion relation between resources (for instance, a machine is included in a cell). A resource has its own calendar and a type defining its operating mode.

For instance,

> List of Operators
> - a Calendar,
> - Operator type (operating mode),
>
> List of Machines. For each Machine, one has:
> - mode: fully automated, half-automated, manual
> - a Calendar,
> - Machine type (operating mode),
> - list of authorized Operators
> - number of Operators
>
> List of group of Operators
> - list of Operators belonging to the group
> - number of Operators
>
> List of group of Machines
> - list of Machines belonging to the group
> - number of Machines

Regarding the modelling of the **Product concepts**, we have to define several types of products (for instance, according to the type of TPP) and a main hierarchical structure (Group of Orders, Order, Operation, Detailed Operation).

For instance,

> List of Orders. For each Order o_i, one has:
> - d_i the earliest start date,
> - f_i the latest end date,
> - Order type (operating mode),
> - quantity to perform,
> - priority (from 1 to 5),
> - list (or more generally, graph) of operations,
> - For each operation, one has:
>   - a code and designation,
>   - operation type (operating mode),

- main machine,
- (PD, UD, TD), Preparation, Unit and Transfer Duration,
- minimum size of lots, overlapping quantity between lots,
- list of substitution machines,

2.1.1.2. Elements for Machine Scheduling

To build the Machine Scheduling Algorithm we propose a modelling tool called the Control Module [BER 81]. Each machine has its own Control Module. It contains the part of the Scheduling Algorithm which concerns the machine. It allows to define a **distributed** algorithm which is more robust when the structure of the model is changed (for instance, number of machines) and easy to program on parallel computer [VEE.93].

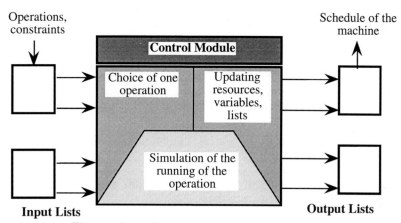

Figure 5 : Structure of a Control Module

A Control Module owns two sets of lists:

- Input Lists,
- Output Lists.

Input Lists are used to manage the operations involving the machine. They may contain the waiting operations with their date of arrival in front of the machine. They may also contain the constrained operations with the date when they should be scheduled.

Output Lists record the history of the machine scheduling. The main output list contains the current scheduling solution. The others allow the algorithm to mark the other attempted solutions and the Decision Points where it is possible to branch or to backtrack.

The Control Module operation corresponds to a three step treatment:

- Choosing an operation according to a rule,
- Simulating the operation,
- Up-dating the state variables and the Input Lists and recording the scheduled operation in the Output Lists.

The Control Module is one of the basic concept used in SIPA+ software to implement the scheduling algorithm [BER 89].

2.1.1.3.   Added Elements for Detailed Scheduling

In this part of the model, more detailed elements are described which allow to simulate precisely:
- the transfer system,
- the tool system,
- the running of a machine-type or operation-type.

The transfer system is modelled separately. Its aim is to validate TD, the Transfer Duration. The Machine Scheduling gives a list of transfers to perform between machines. (Mi, Mj, d, f) means that you have to transfer a product from Machine i to Machine j starting at date d and before date f. The objective of this transfer model is to find a solution to transfer all the products and to meet the d and f dates. If this is not possible, it modifies the dates and transmits the delays to the Machine Scheduling .

The tool system is also modelled separately. Its aim is to do a first validation of PD, the Preparation Duration. This model describes tools transfers and machine configuration globally.

The second validation of PD and sometimes UD is done by detailed simulation of the operations on the machine. For this, we describe the list of Operations. For each Operation op_i, one has :
- d_i the starting date and f_i the latest ending date calculated by Machine Scheduling,
- related order o_i,
- list (or more generally, graph) of previous and next operations,
- For each operation, one has :
    - a code and designation,
    - operation type (operating mode),
    - main machine,
    - UD, Unit Duration,
    - minimum size of lots, overlapping quantity between lots,
    - list of substitution machines.

The operating modes are described with Petri nets [COU.89]. They describe precisely the various states of the objects involved (machine, robot, pallet, conveyor, part, data,...).

2.1.2.   Predictive Control Modelling

Predictive Control concerns Planning Activity, whose running mode is periodic. This FMS Planning Activity is nested inside a structured hierarchy of Planning Activities. In order to integrate the FMS in the whole Production System properly, the FMS Planning Activity must be defined in coherence with the general Planning Function.

We will first explain the various Hierarchical Planning Strategies and their main concepts: horizon, period, Decision Frame. We will then show the FMS Planning Strategy inside this hierarchical structure.

2.1.2.1. Hierarchical planning

This is a classical method in Production Control. It consists in choosing a detail level to define tasks and data according to a given planning horizon [POR.87]. Four planning levels are obtained from Long Term to Very Short Term. This decomposition is very often used in industry [DOU.84].

Other approaches exist, where planning decisions are not organized in a purely hierarchical structure. [JON.90] proposes a combination of two structures, one hierarchical, one heterarchical.

In this decomposition, planning decisions at each level are considered as constraints and objectives for lower levels. Each level has a distinct couple (horizon, period) [DOU.84]. As a general rule, horizon and period decrease as one approaches toward lower levels.

In order to realise a plan, it is necessary to define a co-operation and a synchronisation between levels, to run their planning activity coherently. Two solving strategies are considered:
- Top-down approach,
- Integrally co-ordinated approach.

We propose an intermediate approach: the Partially co-ordinated approach [ARC.91-1]. To realise this approach, one must know how to determine the part of the plan to transmit through the lower levels. In this aim, we define the State of a planning activity and the Decision Frame for a planning activity.

a)   The Top-down approach

The Top-down approach's aim is to calculate a plan through a single Top-down pass. A plan elaborated at one level is transmitted to the lower level as a set of absolute objectives and constraints.

The reliability of this approach depends on the capability for a level to know what can be made on the lower level. For this, the work load of the controlled system is calculated but for the moment we do not know exactly how to do such an evaluation. Because that depends on the set of o_j and their related TPP. So, although this approach is not time-consuming, it leads to sub- or over-evaluation of the shop capacity.

b)   The Integrally co-ordinated approach

The Integrally co-ordinated approach's aim is to calculate a plan through dialogues between adjacent levels. The number of passes is not limited, and each level may contest the objectives and constraints coming from the upper level. The objective is to be sure that the planning solution elaborated is fully admissible.

The main objection to this approach is that it is time-consuming. It is difficult to evaluate the minimum amount of time required to find a first solution, because the lowest level may oblige to re-think the whole hierarchical process planning.

But this method has the advantage of giving a very good solution. Therefore, it seems interesting to be able to use it but if you have time.

210

c)   *The Partially co-ordinated approach*

The Partially co-ordinated approach tries to make a compromise between the two former methods. The objective is to combine the rapidity of the first method and the quality of the second one. For this, we define the **stage** notion. It is one step in the solving hierarchical process.

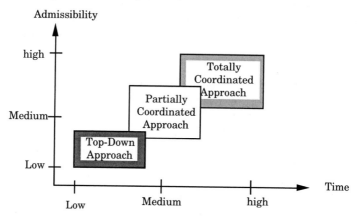

Figure 6 : Hierarchical Planning Approach.

A stage is defined in associating two adjacent levels to find a plan satisfying both levels. The control is then given to the next lower stage, without a possibility to return.

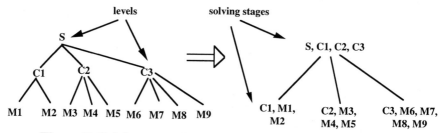

Figure 7 : Solving stage definition with a three-level hierarchy (Shop, Cell, Machine).

On one stage, one has one entity on the upper level, and n on the lower level. The upper entity elaborates a plan and asks the n entities to approve or disapprove of its plan. This dialogue is stopped when the two levels agree or when the maximum number of passes is reached.

d)   *State of a planning activity*

In the GRAI Method, the planning activity has a periodic execution. A decision level L_i is characterised by a horizon, H_i and a period, P_i. The planning activity (PA_i) is activated every period P_i. An activable planning activity (PA_i) may be in one of the two following states:

- the <u>auto-activable state,</u> if and only if the planning activity PA_{i+1} of level L_{i+1} (H_{i+1}, P_{i+1}) has not reached the end of its period,

- the <u>waiting state,</u> to be activated by the planning activity PA_{i+1}, if and only if the planning activity PA_{i+1} has not reached the end of its period P_{i+1}.

e)   *Decision Frame for a planning activity*

Let two adjacent planning activities be PA_{i+1} and PA_i, we called the data they exchanged Decision Frame for a planning activity. This frame consists of:

- the products to plan,
- the TPP,
- **d_i** the earliest starting date and **f_i** the latest ending date,
- the number of the current pass,
- the maximum number of passes.

We called Hc_i the Horizon of the Decision Frame for a planning activity. It is the part of the H_{i+1} whose data has to be transmitted by level L_{i+1} to level L_i.

Hc_i is bounded by H_i and $H_i + P_{i+1}$ because PA_i needs information on the H_i horizon to make a plan.

$$H_i \; < \; H_{Ci} \; < \; P_{i+1} + H_i$$

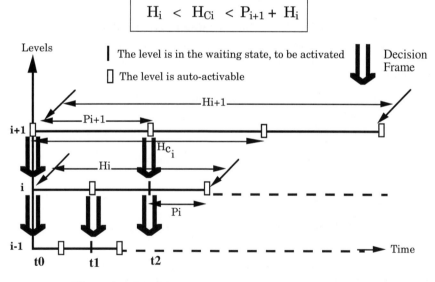

Figure 8 : Decision Frame for a planning activity.

AP_i runs every period P_i. In t_2, AP_i receives a new frame from AP_{i+1}. So, the last time AP_i needs frame information to make a plan before t_2, is t_1. Then, we can write:

$Hc_i = (t_2 - t_0) + H_i$, i.e.:

$$H_{Ci} \; = \; H_i + P_{i+1} - P_i$$

212

Let $\left(H_i = k_i.P_i \ , \ H_{i+1} = k_{i+1}.P_{i+1}\right)$, We obtain the following relations:

Relation between P_{i+1}, P_i and Hc_i

$$\boxed{H_{Ci} = P_{i+1} - \left(1 - k_i\right) \ P_i}$$

Relation between H_{i+1}, H_i and Hc_i

$$\boxed{H_{Ci} = \left(\frac{1}{k_{i+1}}\right) * H_{i+1} - \left(\frac{1 - k_i}{k_i}\right) * H_i}$$

We have determined the part of the plan to be transmitted downto the lower level. Now, we have to state clearly how FMS Control dialogs with MTMS to elaborate the Reference Plan.

2.1.2.2. Principles and scheme of a hierarchical planning strategy.

Viewed from the FMS the partially co-ordinated approach is a dialogue between MTMS and the Load Evaluation and Machines Affectation task. The Decision Frame sent by the MTMS Planning activity is decomposed with TPP information, and the machine load is smoothed out. Load smoothing is done by means of Machines Affection. These affectations are obtained through algorithms and by a man-machine co-operation through a convivial interface. We design and develop such an interface in [BER 90]. An example of a screen is presented in § 3.5.2. If Flexibility Knowledge is specified, it is also possible to make these choices through an inference engine (§ 3.3.1.1).

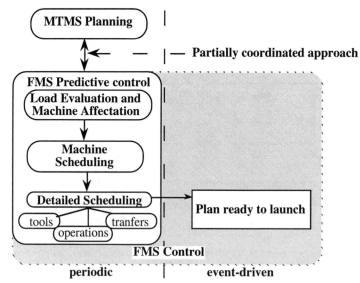

Figure 9 : Decision Frame for a planning activity

Having the machine scheduling task run inside a limited amount of time is the main objective to reach, even if some of the date objectives are not respected. The reason is that if no Reference Plan is elaborated the FMS Control is blind. This is why we suggest to use a List Method (for its rapidity) and an improving process based on algorithms and Man-Machine Interactions both. The improving algorithm is based on permutations made by a Simulated Annealing algorithm [ELO. 92].

For the machine scheduling task, the manufacturing operations are modelled by means of three durations:

> - PD, Preparation Duration, the amount of time required to prepare the machine for a lot of parts,
> - UD, Unit Duration, the amount of time required to process one part,
> - TD, Transfer Duration, the amount of time required to transfer the part to the next machine.

In the FMS, these durations are not known with the same precision. Generally, UD is the best known duration because it is related to a fully automated process. PD is often well-known, but if a tool is shared with another machine PD may vary. And TD is the less well-known duration, because it depends on the work load of the transfer system.

The machine scheduling task elaborates a first schedule, using these three durations. Then, the detailed scheduling task makes a Discrete-Event Simulation to check if PD and TD durations have been evaluated correctly. If necessary, the PD and TD durations are increased, and the schedule is adjusted. We use the SIMGRAI tool to perform this task [BER 85]. [MON.90] proposes an interesting analysis of dispatching rules and how to select them with a discrete event simulator. [MUK.91] has developed an integrated model to realize jointly tool allocation task and sheduling tasks for parts, pallets, machines and conveyors.

2.1.3. Modelling the Auscultation and Diagnosis of disturbances by Flow-Profiles

The auscultation-diagnosis method that we will suggest is based on the use of FLOW-PROFILES [ARC.91-2]. These flow-profiles establish a link between the theoretical production program and the actual manufacturing situations. Other similar methods are proposed, see [WAR.91] and [BOC.91].

The problem of auscultation is basically a problem of reduction of the set of points to be observed (tested) when a dysfunctioning symptom appears. In simple cases, such as machine or tool failures, the point to be observed is directly indicated by the location of the symptom. In complex situations, such as bottlenecks or important manufacturing delay (the symptoms), it is necessary to check every preceding point in the flow of parts and in the circuit of machines. Thus, the number of points to be observed may become very large and incompatible with the production program. The flow-profiles will allow, first of all, to locate the hidden causes of the symptoms, and then, to reduce the number of points to be observed.

A flow-profile is a 3-attribute diagram indicating the load situations of a given machine with respect to time (figure 10) :

Pf = (load L, Waiting Time WT, colour C)

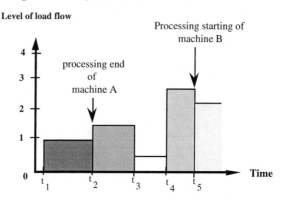

Figure 10 : Flow-profile

Let this machine be called machine B; the load may come from another machine, called A. Load-level L indicates the amount of work that the machine B is going to undertake. Waiting time WT indicates the period during which the intermediate stock between machines B and A remains constant. Colour C indicates the type of parts on which the machine B will operate. It indicates in fact one type of operations, or one group of operations. With respect to the auscultation problem, the aim of the flow-profiles is to reflect various transfer situations between machines. For instance :

(a) At t1, end of one processing by machine A

(b) Between t1 and t2, the intermediate stock remains constant

(c) At t2, end of another processing by machine A, hence, an increase of the waiting stock.

(d) At t3, machine B starts to work, therefore, a decrease of the intermediate stock level.

(e) Between t3 and t4, the intermediate stock remains constant

(f) At t4, more work comes from machine A

The diagram is related to the pair of machines A and B. Should machine B receive work from several other machines A1, A2, A3,..., one has to establish

- firstly, several flow-profiles A1-B, A2-B, A3-B, ...

- secondly, one global flow-profile by summing up the individual ones.

For auscultation and diagnosis, we need a Reference Flow-Profiles (RFP) which reflects the given production program (Reference Plan). One should be aware that the theoretical performances of the machines, as assumed in the production program, may be different in reality. Therefore, this RFP must be permanently, (or periodically), adjusted according to the real situation.

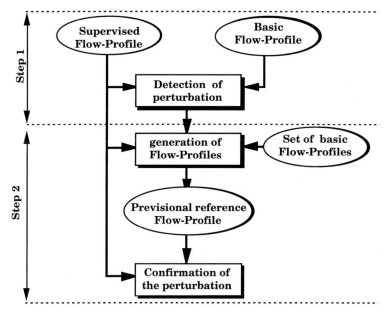

Figure 11 : Detection process

Practically, RFP use for auscultation and diagnosis proceeds as follows :

step 1   - Occurrence of a significant symptom at time t_n
step 2   - Determination of the RFP, the part from t_0 to t_n
step 3   - Determination of the actual flow-profile which is different from the RFP (even up-dated) (The RFP reflects situations which are within the "Earliest-Latest" margin of the operations. Any failure-Symptom indicates that some operations are on the fringe or even beyond this margin)

2.1.4. Modelling the Therapy by Manufacturing Environment Multi-graph

The modelling approach may be summarized as follows. From the notion of Reduced Manufacturing Environment we define a Multi-graph whose edges are the predicted path of the products, and whose nodes are the machines. Using connex graphs, we define the Minimal Potential-Influence Zone of the plan after a disturbance has occurred.

2.1.4.1. Manufacturing Environment of a General Workshop (MEGW)

For a given [0,T] horizon, the MEGW is defined as the set of usable elements (machines and equipment) and :

- products to be manufactured

- TPP and sub-TPP associated with these products

- physical configurations of the manufacturing elements

2.1.4.2. Reduced Manufacturing Environment (Ev)

For a given horizon [0,T], the Ev is defined under the following conditions:

- Products associated with <u>linear</u> TPP representing the programmed and non-executed operations
- Feasible machines with non-blocking resource supports.

2.1.4.3. Multi-graph of a Reduced Manufacturing Environment Ev

The multi-graph is defined as : G=(S,A), where :

<u>S:</u> is the set of nodes associated to machines and equipment, defined by the following bijective function

$$f: M \rightarrow S \text{ verifying the conditions}$$

$$\forall (m_i, m_k) \in M^2 \ f(m_i) \neq f(m_k)$$

$$\forall m_i \in M \ \exists s_i \in S / f(m_i) = s_i$$

<u>A:</u> is the set of edges such as we construct an edge between s_i and s_{i+1}, if there exists a product whose predicted path in the workshop transits in either direction between m_i and m_{i+1}. Between s_i and s_{i+1}, for a given product, there will be as many edges as transitions predicted between the 2 associated machines.

For a given workshop structure: $m_1, m_2,...,m_{i-1}, m_i, m_{i+1},...,m_k$; for a given product p_i associated with a linear TPP g_i, we construct a <u>sub-Graph</u> with k-1 links defined in the following way:

$\forall$ i=1,···,k-1, we connect with one edge the nodes si, si+1 associated with machines m_i and m_{i+1}

The Multi-graph of the manufacturing environment of a given workshop is the <u>union</u> of the set of sub-graphs associated to the products of the manufacturing environment.

<u>Example</u>

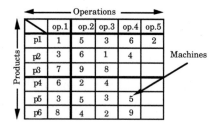

	op.1	op.2	op.3	op.4	op.5
p1	1	5	3	6	2
p2	3	6	1	4	
p3	7	9	8		
p4	6	2	4		
p5	3	5	3	5	
p6	8	4	2	9	

Figure 12 : Manufacturing process.

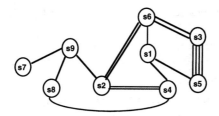

Figure 13 : Associated multi-graph.

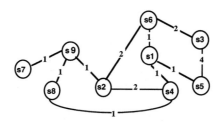

Figure 14 : Associated valued multi-graph.

Let us consider a production program in which :

- six products (p_1 to p_6) are involved
- each product may involve1 to 5 operations ($op1$ to $op5$)
- 9 machines are used: (1 to 9)

Using the TPP associated to the products, the scheduling task assigns a machine to each operation for each product. The schedule-table (figure 12) is obtained.

From the schedule-table, we deduce the associated multi-graph of figure 13:

- each node represents the situation of a machine (s_1 to s_9 correspond to the 9 machines),
- there is one edge (non-oriented link) each time there is an intermediate succession of manufacturing sequences either of the same product, or between two different products (example: edge between s_8 and s_9).
- there are as many edges as immediate successions (example: 2 edges between s_6 and s_2).

In some production programs, especially when there are few machines (multiple-purpose machines and conveying robots), the associate multi-graph becomes heavy and difficult to manage. In such a case, the multi-graph of figure 13 is replaced by the valued multi-graph of figure 14 where the multi-edge is replaced by a single-edge associated to a number representing the number of edges.

2.1.4.4. Absorbing Capacity of an operation

In a manufacturing schedule, there are often gaps. A gap occurs either because a machine is waiting or because a product is waiting. This gap can be defined with respect to an operation, by considering the consecutive operations of this operation. There are two types of consecutive operations: (figure 15)

- operation Sm(i,j) linked by the machine
- toperation (i+1,j) linked by the product.

For the therapy, these gaps can be used for dividing adjusting reactions.

The Absorbing Capacity $C_A(i,j)$ of an operation (i,j) can therefore be defined as the maximum margin which can be employed to annihilate a perturbation on this operation without disturbing the consecutive operations. The $C_A(i,j)$ can be calculated in the following way.

In the following expressions, the symbols **dd**, **df**, **de** stand for:

dd (i,j) starting date of operation (i,j),
df (i,j) ending date of operation (i,j),
de (i) due date of product i,

Let : $\Delta T_P(i,j)$ be the margin between operation (i,j) and its successor (i+1,j)

- if the operation (i+1,j) exists, then $\Delta T_P(i,j) = dd(i+1,j) - df(i,j)$

- if the operation (i+1,j) does not exist, then $\Delta T_P(i,j) = Max(0, de(j) - df(i,j))$.

Let : $\Delta T_M(i,j)$ be the margin between operation (i,j) and its successor Sm(i,j)

- if operation Sm(i,j) exists, then $\Delta T_M(i,j) = dd(Sm(i,j)) - df(i,j)$

- if operation Sm(i,j) does not exist, then $\Delta T_M(i,j) = \infty$.

The local absorbing capacity of operation (i,j) is then:

$$C_A(i,j)= Min(\Delta T_P(i,j), \Delta T_M(i,j))$$

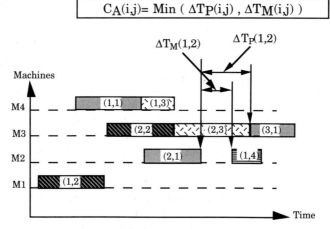

Figure 15 : Margins of an operation (i,j)

2.1.4.5. Minimal Potential-Influence Zone (MPIZ)

One of the key-ideas for solving the NP-complexity scheduling problems, is to decompose the NP-complexity into isolated small ($n_r p_k$)-complexities ([BUR.86], [POR.88]).

If this idea is applied to our method, we need to decompose the Manufacturing Environment of the workshop into small manufacturing environments. Thus , the MPIZ is defined as follows (figure 16) :

            - the zone is initiated by a perturbation,
            - the zone has a spatial range in machines and equipment,
            - the zone has a temporal range,
            - any adjusting reaction in this zone has a minimum influence on the following operations.

2.1.4.6. Connex Graphs

Local zones ($n_r p_k$) reflect the situations of the manufacturing capacities of the environments. The connex graphs are operational tools, which transform zones ($n_r p_k$) into a model, from which the monitor can calculate his adjusting reactions quantitatively.

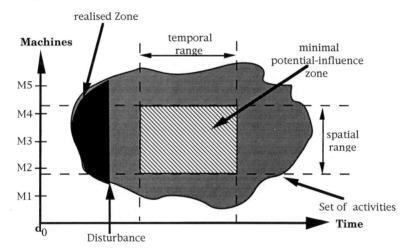

Figure 16 : Minimal potential-influence zone.

A connex graph is a sub-graph of the multi-graph representing the manufacturing environment. The initial multi-graph may contain isolated connex-graphs (connex components of the multi-graph).

A connex component may appear after the execution of some manufacturing operations (figure 17, figure 18).

	op.1	op.2	op.3	op.4	op.5
p1	1	5	3	6	2
p2	3	6	1	4	
p3	7	9	8		
p4	6	2	4		
p5	3	5	3	5	

Figure 17 : New schedule-table.

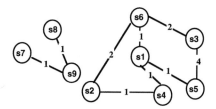

Figure 18 : Valued multi-graph associated to the schedule-table of figure 17.

Example: Let us start with the schedule-table in figure 12. Several manufacturing operations have been executed so that the product p_5 is achieved. Several machine-assignments have occurred, resulting in the schedule-table in the figure 17. The multi-graph in figure 13 becomes then that of figure 18. The latter now contains two connex components. One on the left, and the other one on the right.

2.2. Planning or Predictive Control

2.2.1. Convivial planning

The first principle to respect when specifying a Convivial Planning Software is that the Human Operator responsible for the scheduling task must not be idle in front of his computer, or the Planning Software is not convivial and the Human Operator does not use it.

Secondly, the Human Operator must be able to enter his solutions or constraints easily.

Thirdly, the Convivial Planning Software must explain the solutions it has found.

So, the Convivial Planning Software relies on [BER 87]:

> - an Interactive Man-Machine Interface,
> - a very quick basic algorithm,
> - several improving algorithms running in the background, which can explain the solutions found,
> - many utility functions to analyse the scheduling situation.

2.2.2. Man-Machine Interface

The heart of the Man-Machine Interface for a Convivial Planning Software is the Interactive Graphic Representations for Planning, i.e. a Gantt diagram or a Load diagram.

The meaning of Interactive is that the Graphic Representations are the media used by Man and Computer to co-operate towards a solution. The Graphic Representations have to be reactive, i.e. tasks on a diagram must be easily movable with the mouse, and a double-click should open an information window on the selected task.

A Graphic Representation for Planning is a 2D view of a 3D Abstract Object: the planning. The three dimensions of a planning are:

  - **T**, Time used to represent Planning Horizon, Period, operation duration,...
  - **P**, Product, a generic term used for all the materials transformed by the FMS,
  - **R**, Resource, represents all the objects that serve as support for the FMS operation.

To specify a Graphic Representation for Planning, we have to define:

  - a **selection** among these three sets (T, P, R); i.e. an interval of Time, a sub-set of Products and a sub-set of Resources,
  - a **graph type**: Gantt, Load, Day planning,
  - the set of **tools** used to Interact with the planning.

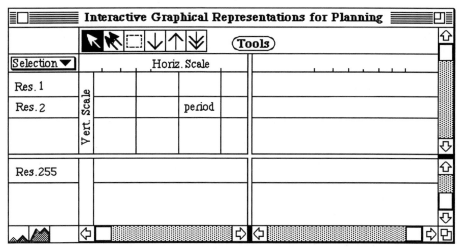

Figure 19 : Proposition of a Planning Interface

2.2.3.   List method

The proposed List Algorithm is based on the use of the [BER.83] Control Module.

H is the horizon, **T0** is the beginning date of the plan, **T** is the current date, the **global algorithm** dispatches operations in the input lists of the Control Modules. Each Control Module with its **local algorithm** manages the choice of operations in its input list.

T is set to **T0**
WHILE T < T0 + H **DO**
    FOR each free machine **DO**
        | Activate **ControlModule** of the Machine at date T
        | Update the Input Lists of others Control_Modules
    END OF FOR
    Increase T
END OF WHILE

Figure 20 : Global algorithm

To Choose an operation **Oi** in the Input List according to a rule,
IF Oi is chosen **THEN**
        | **Simulate** the operation on the Machine at date T
        | **Update** the state variables and the Input Lists and **record**
        the scheduled operation in the Output Lists
END OF IF

Figure 21 : Local algorithm of the Control_Module

This algorithm is very simple in its general structure. This dispatched structure makes it easy to execute by a parallel computer. This is important to reduce the amount of time required for scheduling.

2.2.4. Improving solution with a Simulated Annealing algorithm

The Simulated Annealing algorithm to improve Shop scheduling solutions is interesting because of its generality. Its structure is not dependent on the structure of the Shop.

The first task to perform is to choose an evaluation criterion to measure the quality of the scheduling solution. This criterion may be used :

    - to minimise the total processing time,
    - to minimise the total processing time of a set of manufacturing orders (for example, those with a high priority),
    - to maximise the load of bottlenecked machines.

The Simulated Annealing algorithm [BON. 91] is as follows :

Begin
    To evaluate criterion C for the current scheduling solution
    k <-- 0 (initialisation)
    To evaluate starting temperature T(0) (T is temperature)
    While (the stop criterion is not verified) **Do**
    Begin
        Repeat_Count <-- 0 (initialisation)
        While (Repeat_Count < N(k)) **Do**
        Begin
            Change a product choice for one machine

Make a new schedule
To evaluate criterion C' for the new scheduling solution
If (criterion value C' better than C) **Then**
    the new scheduling solution becomes the current one
Else
    A(T) = exp ((C - C') / T)
    a random number p is generated in the range [0, 1]
    If p < A(T) **Then**
        the new scheduling solution becomes the current one
    Else
        the product choice Change is cancelled
    End of If
End of If
Repeat_Count <- Repeat_Count + 1
End of while
k <-- k + 1
To evaluate T(k)
End of while
End

Figure 22 : The Simulated Annealing algorithm

Temperature T and probability P aim to allow the criterion C to grow between two decreases. Temperature T is a decreasing geometric function T(k) = a.T(k-1); "a" is a constant strictly inferior to 1. The number of repetitions "N(k)" is linked to the maximum number of attempts accepted between two decreases of the criteria; it is often constant. The main loop depends on a stop criterion for the improving process. For us, this stop criterion depends on the amount of time allowed by the user to the computer to devise this improving process.

The Simulated Annealing algorithm is based on the following parameters : the initial schedule and temperature, the decreasing geometric function, the stop criterion and the rule to make the change of product choice.

2.3. Supervising or On-line Control.

The On-line Control is the part of the FMS Control that is most developed in industry. Few generic studies about it are available. The modelling tools commonly used are :

- SADT to structure the physical elements, mainly resources whose work load is limited (machines, equipment, materials, stocks) [KER.93], [BOW.91],
- GRAI-nets [PUN 83], [PUN 84]:
  - to model the activities and the situations of the product flows,
  - to model the command signals and their reception points,
  - to model of the supervising points in the product flows.

To make real-time scheduling decisions, the most common approach is to use dispatching rules [HUT.91], [MAH.90], [MON.90].

2.4. Dynamic Control.

We have presented above (§ 2.2.3 and 2.2.4) the modelling tools to be used :

- flow-profile modelling to reflect manufacturing situations (manufacturing load),
- capacity-diagram modelling to reflect and understand the situations of the capacity utilisation.

We will now explain the procedures to solve the following dynamic problems :

- auscultation and diagnosis,
- therapy.

2.4.1. Methodology for Auscultation and Diagnosis

In terms of FMS planning and control, the auscultation process occurs when the causes of the perturbationsymptom cannot be clearly detected. Once the causes are detected the diagnosing process takes place. In terms of "supervising" and "dynamic adjusting" activities, the diagnosing process is lies between these two activities. This process, in fact, consists of two activities :

(a) perception, which interprets the results of the auscultation, namely, what the causes of the dysfunctioning are; and (b) determination of a field of possible actions which will guide the therapy.

In our method, both auscultation and diagnosis are based on the use of the Situational Diagram of the flow_Profiles of the Workshop machines (Figure 23).

Let :

SIT-P : situation of a machine which produces a load

SIT-C : situation of a machine which receives and consumes a load

n : the number of machines used in a workshop.

The Situational Diagram shows relations between various situations of Flow-Profiles between the machine-situations :

(a) Level 0 : Situation of Flow-Profiles of n machines in SIT-P and n machines in SIT-C (top part of the diagram)

(b) Level 1 : Two situations of Flow-Profiles

(b1) Situation of Flow-Profiles of n machines in SIT-P and the whole of the n machines in SIT-C considered as one machine (middle left)

(b2) Situation of Flow-Profiles of n machines in SIT-C and the whole of the n machines in SIT-P considered as one machine (middle right)

(c) Level 2 : Situation of Flow-Profiles of the situation (b1) plus (b2) (bottom of the diagram)

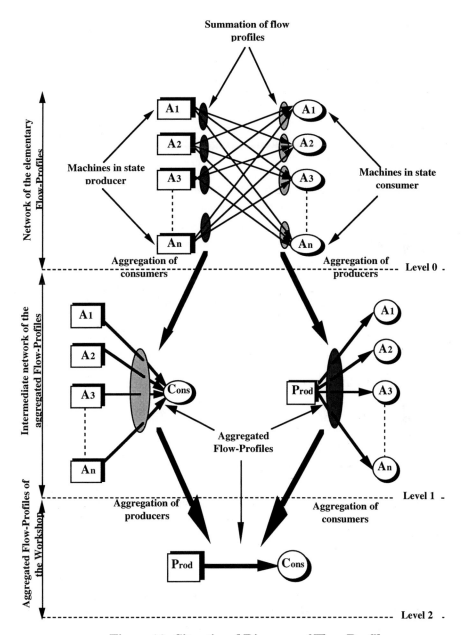

Figure 23 : Situational Diagram of Flow-Profiles

The Flow-Profiles reflect the situations of the manufacturing operations. It is important to note that, theoretically, when there are n machines in the manufacturing process, the total number of Flow-Profiles is n*n (100 only for n = 10). This is why auscultation is difficult. However, for a given production program, this number is much smaller. If each machine is used once, this number is 9 for n = 10. This number remains small even if some of the machines are to be used several times.

The global auscultation and diagnosing process comprises three main steps (each one including several sub-steps) : (Figure 24).

- determination of the observable critical point
- identification of the critical activity
- identification of the faulty actor or actors.

Step 1- Determination of the observable critical point

For this, we use the Situational Diagram of Flow-Profiles (Figure 2). The process starts when a significant perturbation is detected.

Step 1.1 - Evaluation of the tendency

The "tendency" information accelerates the search for the observable critical point. This information is obtained by comparing the reference Flow-profiles (up-dated) and the perturbed one of level 2. An uphill tendency (some machines PRODuce too much load) indicates that the perturbation is on the "PROD" side. A downstream tendency (some machines have too much load to CONSume) indicates that the perturbation is on the "CONS" side.

Step 1.2- Determination of the first machine

At this sub-step, we compare the Reference and the perturbed Flow-profiles at the level 1. When the tendency is "PROD", we identify as the first machine, the perturbed producer. When the tendency is "CONS" we identify as the first machine, the perturbed consumer.

Step 1.3- Determination of the second machine

At this sub-step, we compare the Reference and the Perturbed Flow-Profiles at Level 0, which contains all the elementary flow-profiles. Note that, from Sub-step 1.2, only one producer or one consumer is identified. Therefore, at Sub-step 1.3, according to whether this is a producer or a consumer, we only compare one set of n corresponding consumers or of n corresponding producers.

The final result of Step 1 is one flow-profile connecting a pair of well-defined producer and consumer. The search for the observable critical point requires : 1 comparison at level 2, n comparisons at levels 1 and 0. We therefore reduce a problem of complexity $O(n^2)$ to one of complexity of $O(2n+1)$.

Step 2. Identification of the critical activity

Once the observable critical point is located, we identify the critical activity in space and in time. This is done by using the load diagram corresponding to the flow-profile around the critical point. The various types of causes are identified by comparing the reference load diagram **fp** (defined by the production program) with the real actual load diagram **fr** (Figure 4). The difference between the two diagrams may appear in the length **L** of the waiting time, in the load level **N**, or in the Colour **C** (p : index for reference production program, r : real current situation, t_i : moment in time when the difference is observed).

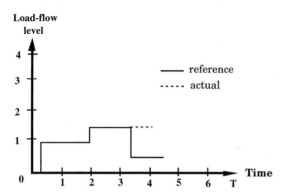

Figure 24 : Comparison of load diagrams

(a) The lengths are different : $L_p^i \neq L_r^i$
Furthermore, load-levels are also different :

   (a1) $L_p^i < L_r^i$ and $N_p^{i+1} > N_p^i$
   The perturbation causes may be :
    - machine failure of the uphill unit A
    - under-evaluation of the processing time of A

   (a2) $L_p^i < L_r^i$ and $N_p^{i+1} < N_p^i$
   The causes may be :
    - machine failure of the B downstream unit
    - procurement delay at B because of conveying problems

   (a3) $L_p^i > L_r^i$ and $N_r^{i+1} > N_r^i$
   The causes may be :
    - over-evaluation of the processing time at A
    - capacity of B larger than previously assumed
    - problems in the percentage of waste parts

   (a4) $L_p^i > L_r^i$ and $N_r^{i+1} < N_r^i$
   The causes may be :
    - the transfer between A and B is too fast
    - procurement problems of the machines
    - wrong choice of machines
(b) Diagnosis on the waste-rate

   This happens if $N_p^i \neq N_r^i$ and $C_p^i = C_r^i$. The situation confirms that the underlying activities are critical.
(c) Diagnosis on the sequencing

   This happens if $C_p^i \neq C_r^i$
   The causes may be

- either a misconveying of parts
- or a wrong sequencing of operations

Step 3 Identification of critical actors

Locating the critical activity allows identification of the critical actor or actors. This is done by representing the activities by means of the GRAI-nets (Graphs with Results and Activities Interrelated) [PUN 84]. In this representation, each activity is symbolically characterised by :

- input and output events,
- intellectual supports,
- material supports,
- operator.

The interrelations between activities are always defined by the fact that the support of one activity always comes from the result-event of another activity. Based on the GRAI-nets, the various causes identified at step 3 are symbolically translated into real elements.

2.4.2.   Therapy Procedure

The aim of the procedure is to help the monitor, after the perturbation has occurred, and after the critical elements have been located, to determine adjusting reactions with a minimum of consequences.

Step 1: Initiation

Auscultation and diagnosis have identified the critical elements, causing the pernicious perturbations. The Monitor displays the schedule-table and the Gantt diagram of the manufacturing program. From the Gantt diagram, the Monitor identifies the perturbation date dp (figure 26). Its task is now to identify the MPIZ. For this, he carries out a spatial and temporal decomposition.

Step 2: Spatial Decomposition

The MPIZ is characterised by a spatial and a temporal range. From the schedule-table and the Gantt diagram, the Monitor generates the associated Multi-graph. (An automatic generator must be implemented in the decision-aid system). If the multi-graph possesses 2 or more connex components, the monitor chooses the smallest one. If not, the Monitor carries out a temporal decomposition.

Step 3: Temporal Decomposition

Each time a manufacturing operation is performed, a rupture may occur in some link of the multi-graph. As a consequence, new connex components may occur. There are three types of ruptures:

type a. The operation achieved is associated in the valued graph with a link of a value superior to one. The value decreases by one unit. No rupture in the graph. The connexity remains the same.

type b. The operation achieved is associated in the valued graph with a link of a value of 1 between nodes s_i and s_{i+1}. After this rupture, there exists no circuit between these two nodes. The initial graph splits into two connex components.

type c. The operation achieved is associated in the valued graph with a link of a value of 1 between nodes s_i and s_{i+1}. After this rupture, there still exists some circuit between these two nodes. No new connex components are created. The connexity of the initial graph remains the same.

For the temporal decomposition, the Monitor displays the schedule-table (figure 25) and the corresponding Gantt diagram (figure 26). He simulates the successive achievements of the manufacturing program (automatic pointer to be implemented in the decision-aid system).

Operations →

	op.1	op.2	op.3	op.4	op.5
p1	9	7	8	7	
p2	7	8	9		
p3	8	6	2		
p4	6	2	4		
p5	3	6	1	4	
p6	1	5	3	6	2
p7	3	5	3	5	

Products →   Machines

Figure 25 : Schedule-table under temporal decomposition.

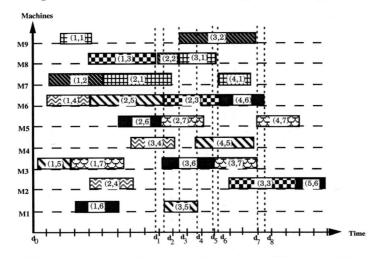

Figure 26 : Gantt diagram under temporal decomposition.

After each operation has been completed, he analyses the new structure of the multi-graph of the manufacturing environment (connection between the pointer and the multi-graph generator to be implemented in the decision-aid system, the latter must have a multi-window possibility). The succession of multi-graphs is shown in figure 27.

The various steps of the decomposition are the following.

a- Initial structure A$_0$ of the multi-graph at date d$_0$.

b- At date d$_1$, operation(1,3) ends. Rupture of link (s$_6$, s$_8$). Graph A$_0$ splits into two connex components A$_1$ and A$_2$.

c- At date d$_2$, operation(2,5) ends. Rupture of link (s$_6$, s$_1$). Sub-graph A$_1$ splits into two connex components A$_{11}$ and A$_{12}$.

d- At date d$_3$, operation(2,2) ends. Rupture of link (s$_9$, s$_8$). Sub-graph A$_2$ splits into two connex components A$_{21}$ and A$_{22}$.

e- At date d$_4$, operation(3,5) ends. Rupture of link (s$_1$, s$_4$). Sub-graph A$_{12}$ splits into two connex components A$_{121}$ and A$_{122}$.

f- At date d$_5$, operation(3,6) ends. Rupture of link (s$_3$, s$_6$). Sub-graph A$_{11}$ splits into two connex components A$_{111}$ and A$_{112}$.

g- At date d$_6$, operation(3,1) ends. Rupture of link (s$_7$, s$_8$). Sub-graph A$_{21}$ splits into two connex components A$_{211}$ and A$_{212}$.

h- At date d$_7$, operation(3,7) ends. Rupture of link (s$_3$, s$_5$). Sub-graph A$_{111}$ splits into two connex components A$_{1111}$ and A$_{1112}$.

i- At date d$_8$, operation(4,6) ends. Rupture of link (s$_2$, s$_6$). Sub-graph A$_{112}$ splits into two connex components A$_{1121}$ and A$_{1122}$.

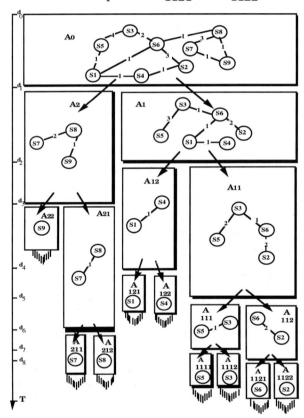

Figure 27 : Temporal decomposition of a multi-graph.

At each step a, b, c, ..., i, the monitor tries to determine the adequate adjusting reactions. If he does not succeed at one particular step, he goes to the next one. The earlier he succeeds with the dates, the better the adjusting reactions are. This procedure is intrinsically an optimising procedure. The minimised criterion are the consequences of the perturbation on the production program.

Step 4: Adjusting through permutation (figure 28)

Once the connex component is identified, the Monitor displays the corresponding schedule-table and Gantt diagram. He tries to absorb the perturbation through permutation. Two operations (i,j) and (k,l) scheduled on the same resource m_i are said to be permutable if

$$\text{Max } (df (i-1,j), df (k-1,l)) < \text{Min } (dd(i,j),dd(k,l))$$
$$\text{Max } (df (i,j),df (k,l)) < \text{Min } (dd(i+1,j), dd(k+1,l))$$

(Automatic computation of these conditions to be implemented in the decision-aid system).

Step 5: Adjusting through transfer (figure 29)

An operation (i,j) scheduled on resource mi is said to be transferable if there exists at least one alternative resource $m_k \neq m_i$ available between dd(i,j) and df(i,j).

(Automatic prospecting to be implemented in the decision-aid system).

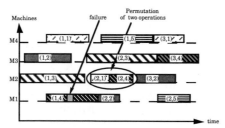

Figure 28 : Permutation of manufacturing operations.

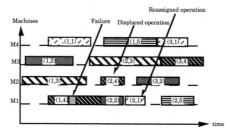

Figure 29 : Transfer of manufacturing operations.

3.   Synthesis and Software architecture

From the tools and methods chosen for solving FMS Control problems this part proposes a methodology to build the Software architecture. The aim is not to describe a trivial programming method but to insist on the difficult points:

> - general principles to organize the global architecture,
> - construction of the various elements of the Software: Database, Knowledge, Programs, Interfaces,
> - language and development methodology.

The FMS Control we propose is a Convivial Control. For this reason, the presentation emphasises the heart of a Convivial Software: the Man-Machine interface. Then the FMS Control has to be integrated in the overall Enterprise Control. So, the software architecture must be designed and developed with a modular organisation [SAN.91].

3.1. General architecture.

3.1.1.   Logical Architecture

The following scheme (figure 30) defines the general architecture of the FMS Planning and Control System for one part of the FMS, i.e. Cell control or Central control in figure 31.

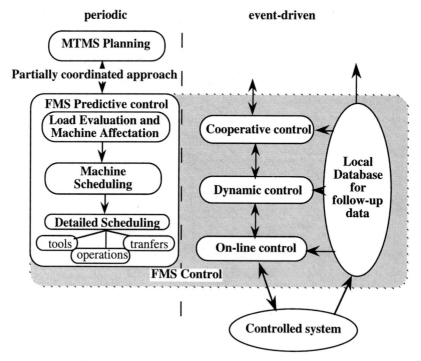

Figure 30 : FMS Planning and Control System.

Co-operative control concerns exchange with upper level to find solution when the Decision frame cannot be respected.

Local Database is used to maintain an image of the controlled system. Some aggregation functions allow to transmit to the upper level the results achieved by the controlled system.

3.1.2.   Physical Architecture

The practical aim of the Physical Architecture is :

> - operators must use the Control Software,
> - operators have to be motivated to obtain the best "score" with their FMS,
> - operators have to be involved in the improvement of functions and interfaces.

For this, the FMS Control Software must be:

> - an agreeable software to use, it must encourage to "play" with all the functions,
> - efficient to obtain information, to print reports on all the aspects of the FMS,
> - a media allowing the decision makers to communicate between them but also   providing all operators with information and knowledge,

So, the actual trend is to install a local network of Workstations :

> - one station for the Central Control,
> - one station for each human cell responsible for a part of the FMS (Cells, Complex machines,...),
> - one station for each function (transfer, database, tools, storage, Direct Numeric Control).

Each workstation has the following specifications :

> - connected to the other workstations by a network,
> - running with multi-task and real-time operating system,
> - owning a multi-windowing interface capable of connecting several colour screens,
> - equipped with office utilities for operators: mail, word processing, spreadsheet and copy/paste functions between office software and control software.

234

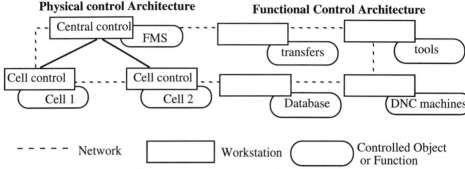

Figure 31 : Physical Architecture.

A number of principles have to be respected when using this physical architecture :

- **Decision-maker code** : each operator liable to use the workstation must have a card or a badge to control access to a number of critical functions of the FMS Control Software,
- **Flexibility in control** : each operator must have access to its tasks from all the Workstations,
- **Access type** : each task or program may be used under different modes (consultation, modifications, information) according to the responsibility of the operator.

3.2. Building of the Database (object-oriented approach).

The Database is a fundamental element of the FMS Control Software. It must be a Real-time Database with the following problems to solve :

- **Access time** : when a disturbance is detected or when a schedule is computed a lot of data has to be reached rapidly.
- **Updating** : all the shared information must be recorded in and retrieved from the Database to prevent inconsistencies between two programs.
- **Multiple domains :** the Database contains several aspects of the same data according to the domain : routes operations for schedules, operations for D.N.C.

In an FMS Control Software, we are essentially concerned with **Access time**. So, we propose to realise a RAM Database through an object language (RAM: Random Access Memory, a work memory whose Access time is quicker than disk memory). This Database will be managed by a dedicated workstation with a great amount of RAM memory to be accessed rapidly from any workstation.

The contents of the Database are essentially:

- the **object models** of the Controlled FMS
    - object model of the FMS (Common kernel),
    - added models: Machine Scheduling, Detailed Scheduling,
    - object model of the Flow-Profiles
    - object model of the Manufacturing Environment Multi-graph

- the **object model** of the FMS Planning Control System

Each program of the global software is an object in this structure. This concept allows to design an Open Control System easy to develop and maintain. The software development task is unified in one programming environment.

3.3. Building of the Knowledge-base (predicate-logic approach).

We will try to specify two points about the building of the Knowledge-base :

- the content of a Knowledge-base for an FMS Control Software,
- the way to enter knowledge in the base.

Two other important questions are not treated in this paper : the management of the knowledge coherence and Knowledge-base organisation to manage jointly object models and sets of rules. For a first response to this research problem, see [PAC 92].

3.3.1. Contents of the Knowledge-base

Generally, the Knowledge-base contains information about how to use data and models in the FMS Control Software. This knowledge may be divided into four parts:

- Knowledge about Flexibility,
- History of the FMS operation,
- Knowledge about Man-Machine Interactions,
- Knowledge about Dynamic Control.

3.3.1.1. Knowledge about Flexibility

In many control situations, the operators or the programs need information about the flexibility of the shop. Two main types of flexibility are considered :

- Machine or equipment Flexibility,
- Product Flexibility.

Each equipment of the FMS has various operating modes. A normal mode is chosen but one may have to switch to another mode in order to solve some disturbance situations. This technical information is often unknown to the Control Software. It is important to describe all these modes and to enter them in the Knowledge-base. Each machine may also have several configurations. These configurations are related to the tools, the programs or specific equipment attached to the machine. If a formal description of these configurations is available in the Knowledge-base, the FMS Control Software (or the operator) will be able to search a substitution solution when a disturbance occurs.

To meet a manufacturing order, several TPP may be used to obtain one same product. In the same way, within a route, for certain operations, several machines may be used. This flexibility knowledge about the product should be described to give the operators and the software a capability to react and adjust the plan.

All these possible choices (operating mode, machine configuration, route,...) may be guided (or inferred) by an Analysis of the set of manufacturing orders sent by the MTMS. This analysis may be made by a rule-based expert system [SUN.89].

3.3.1.2.   History

This part of the Knowledge-base concerns the record of the FMS operation and its Control Software. The history of the shop allows operators or software to compare present situations with past situations. We propose to classify this historical knowledge into four parts:

- Disturbance situations,
- Strategies used to solve disturbances situations,
- AI solving strategies for generic situations,
- Statistics about each route type.

3.3.1.3.   Knowledge about Man-Machine Interactions

This part of the Knowledge-base is concerned with the manner to introduce situations to the Decision-Makers. In a disturbance situation, the FMS Control Software must show the operator the identified situation and the result of the Dynamic Control analysis (Auscultation, Diagnosis and Therapy). But, according to the disturbance situation and to the operator, the Man-Machine Interactions are different. So, the FMS Control Software must adapt its interface and must learn new ways of presentation.

3.3.1.4.   Knowledge about Dynamic Control

This part of the Knowledge-base is concerned with how to use the Flow-Profiles and Manufacturing Environment Multi-graph tools to solve disturbance problems [ARC.91-3].

3.3.2.   Principle of Human know-how and knowledge integration

Formalising Human know-how is difficult, which makes it necessary to adopt a pragmatic approach. Today, in FMS Control, we are not capable of programming the computer learning of Human know-how. So, we propose a progressive integration by defining Software Intelligence Levels (Esprit Project n°418 [Milin 87]).

The principle of this approach is to integrate gradually Human know-how by observing Man-Machine Interactions. This allows to develop the software step by step without re-programming the lower levels.

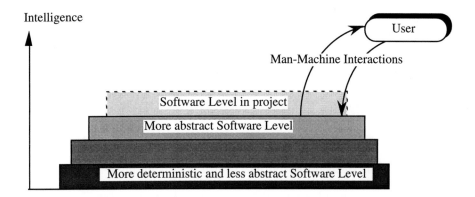

Figure 32: Integration of Human know-how in Software Intelligence Levels

The intelligence of a Software Level is related to the lower levels already implemented. In such a structure, the capabilities of the new level become simple functions of the new software. Once implemented, they are not intelligent any more. Then, Intelligence is in the user and in the way to use these new functions. Accordingly, we have to analyse how this is done and try to design the new software level.

So, in order to integrate knowledge in the FMS Control Software, we propose the following steps :

- Observe and record the FMS Control Software operation including the Man-Machine Interactions;
- Analyse the previous records to define new concepts in the Knowledge-base;
- Design a Man-Machine Interface to allow operators to use this Knowledge-base convivially;
- Observe and record the Man-Machine Interactions on this new interface to identify generic behaviours;
- Design and program new functions using the new concepts of the Knowledge-base.

3.4. Building the processing modules.

Each processing module is a instance of an object Application.

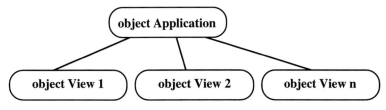

Figure 33 : Object structure of a program.

An application is a task or a computer job. It manages several views. A view is generally located in a window.

The skeleton of the object Applications depends on the Graphic Environment used (X-Windows, Windows, Motif, OpenLook, Macintosh,...). Programmers should refer to the related "toolbox".

3.5. Building Man-Machine interfaces.

Before programming interfaces, one should analyse the Man-Machine Interactions precisely [BOY 92]. This can be done by the Scenario Method based on the analysis of Decision Activities, whose operating mode depends both on Man and Machine through an interface. When the scenarios are defined, the interface screens or windows must be specified. Each interface is an instance of the object View. Programming a Man-Machine interface is very expensive. That is why, it is important, for the work to be reusable, to build software components independently of computer languages and machines [NEE.90]. For this, Graphic Abstract Data Type is a good tool to design the more difficult part of an interface: namely the Interactive Diagram.

3.5.1. Scenarios Method to define Man-Machine Interactions.

This part presents an analysis and design method for Interactive Decision Aid Systems using advanced Man-Machine Interfaces (pull-down menus, multi-windowing,...) [BER 90]. The basic principle is the same as in the GRAI's method: decision activities are first analysed, then those supported by Man and Machine both are detailed. To build a scenario it is necessary to identify Man actions and Machine actions and therefore the events related to the synchronisation of these actions.

First, we will present a form of GRAI net adapted to the modelling of Man-Machine Decision Activities. Then, the second tool is the scenario using a three columns table for stating Man-Machine Interactions and Events.

3.5.1.1. Man-Machine Decision Activities

Man reasoning may be characterised by active phases (Decision Activity) and passive phases (Decisional Situations). These two concepts and the Activity and Event concepts in Discrete Event Modelling are very similar. To describe a reasoning process is to define and relate all these active and passive phases.

The GRAI Net is used to make a global description of the reasoning process as shown in figure 34.

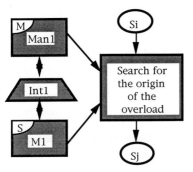

Figure 34 : Decision Activity with Man-Machine Interactions

Each Decision Activity begins with an initial situation **Si** and ends with a final situation **Sj**. Each decisional situation is a step in the reasoning process, characterised by a screen **Ei**. The Decision Activities with Man-Machine Interactions are supported both by Man **Man1** and Machine **M1**. In the upper left-hand corner it is indicated whether the support is the Main or a Secondary support. The Main support manages the Interactive process. Between these two supports, a trapezium formalise the interface **Int1**.

The **Ei** screens characterise the decisional situations and are specified by :

         - the list of active pull-down menus,
         - the list of opened windows and the active window **Fi**,
         - the objects (icons), outside the windows,
         - a copy of the screen in its more classical state.

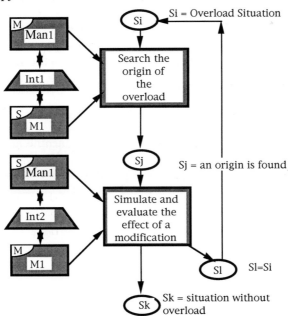

Figure 35 : Decision Process with Man-Machine Interactions

Figure 35 shows an example of Decision Process with Man-Machine Interactions. The objective is to solve an overload problem before scheduling.

This iterative Decision Process consists of two activities:

> - the first activity is concerned with the search for the origin of the overload; it is driven by the man aided by the machine (**H1** is the main support),
> - the second activity aims at modifying the plan and make a simulation to evaluate the consequence of this modification; it is driven by the computer (**M1** is the main support); two situations may be obtained **Sl=Si** where there is still an overload and **Sk** where the overload problem is solved.

Two different interfaces **Int1** and **Int2** support this interaction.

3.5.1.2. Scenario

To define the scenario concept, we have to recall the two main types of Man-Machine Dialogue:

> - the modal dialogue,
> - the non-modal dialogue.

The **modal dialogue** is run under computer control. This dialogue asks a precise information from the human operator. As long as the man has not given his answer within the frame defined by the dialogue, he cannot take another step. This type of dialogue is interesting when there is a risk of data incoherence. This mode is easy to manage (and to program) but it is hardly interactive.

The **non-modal dialogue** is run under man control. He is free to select a menu, to activate a window or any object on the screen. This type of dialogue is highly interactive and allows a software to help a human user efficiently.

A Man-Machine scenario has to characterise these two interaction modes. It is composed of a three-column table:

> - the first column describes man actions (**Aman i**),
> - the second one contains the **events ei** and **screens Scri** characterising the beginning of a man or machine action,
> - the third one defines the machine actions (**Am i**).

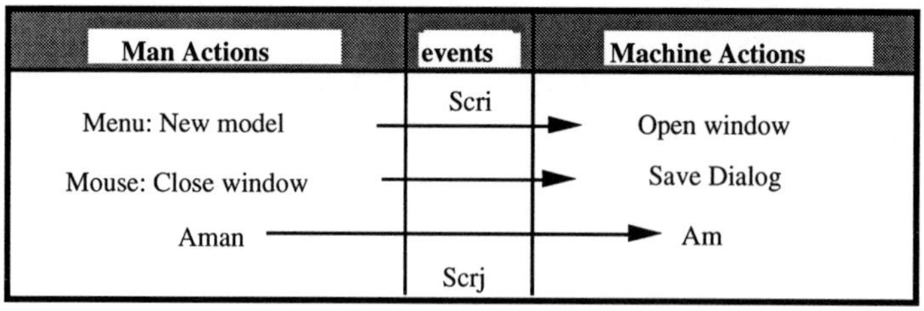

Man Actions	events	Machine Actions
	Scri	
Menu: New model		Open window
Mouse: Close window		Save Dialog
Aman		Am
	Scrj	

Figure 36 : Example of a scenario

Events are symbolized by arrows (figure 36).

Each Decision Activity is generally detailed by a scenario. That is why it begins with a screen **Scri** (initial situation) and ends with a screen **Scrj** (final situation). This example shows that a man action on the menu: New model leads to open a window.

The scenario of figure 37 describes a Decision Activity beginning with a modal dialogue. The initial screen **Scri** leads necessarily to the man action **Aman1** which leads to machine action **Am1**. The end of the Decision Activity is decided by man action **Aman2** which automatically leads to machine action **Am2** and screen **Scrj** (final situation).

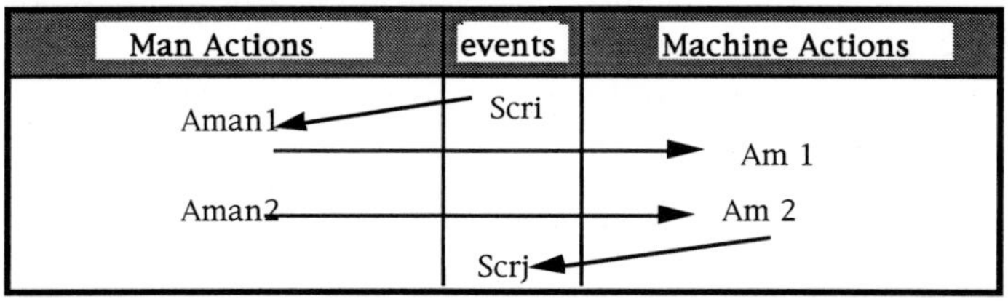

Figure 37 : Scenario with events linked to initial and final situations

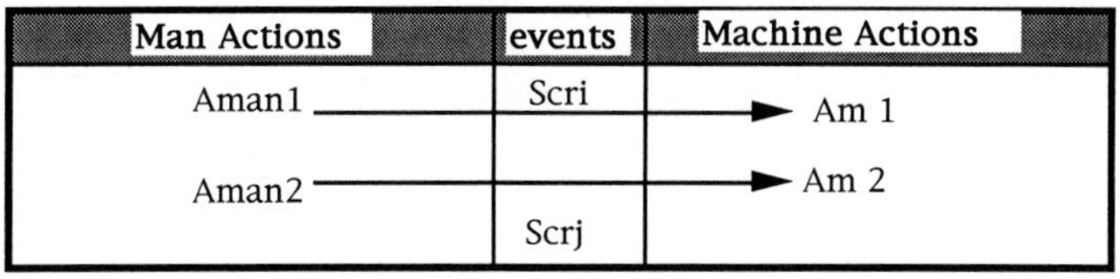

Figure 38 : Scenario with non-modal dialogue

In figure 38, all the arrows of the scenario go from man to machine. No event asks for a compulsory response from man. He can choose freely between **Aman1** or **Aman2** action. This is a purely non-modal scenario.

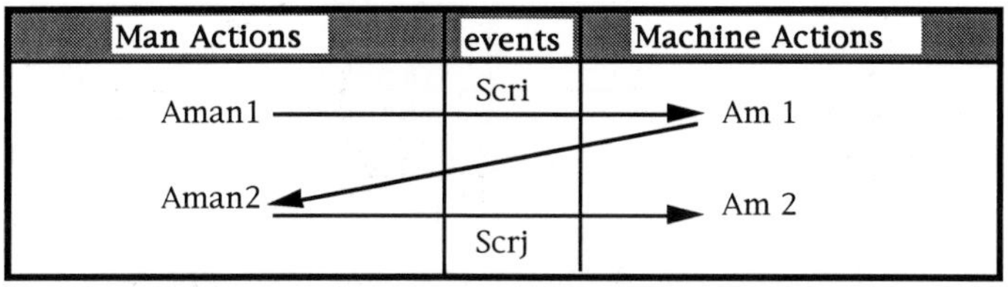

Figure 39 : Scenario with modal dialogue

Figure 39 represents a purely modal scenario. Man action **Aman2** is the consequence of machine action **Am1**.

3.5.2. Graphic Abstract Data Types to specify Interactive Diagrams.

Graphic Abstract Data Types are a way to describe an Interactive Diagram independently of a programming language. The aim is to formalise the Interactive Diagram at two levels:

242

- its geometric form,
- its behaviour.

A Graphic Abstract Data Type is composed of three parts:

- a geometric form and its variables,
- attributes,
- methods describing the dynamic behaviour of the Interactive Diagram.

Here is an example of a geometric form and of some of its variables. Hl and Lc are two variables corresponding respectively to the height of a line and the width of a column. The variables are related to the geometric form of the Interactive Diagram.

If we use this basic Abstract Data Type to define a Resource Gantt Diagram, a line is associated to a resource and a column to a period. These definitions are attributes of the Graphic Abstract Data Type. The attributes are related to the semantic contents of the Interactive Diagram.

The classical methods are:

- creation of a new Interactive Diagram,
- destruction of an Interactive Diagram,
- drawing method,
- access to attributes and variables,
- management of mouse actions, menu actions, keyboard actions...

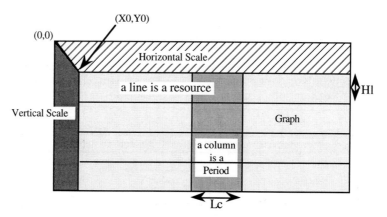

Figure 40 : Geometric form of a Graphic Abstract Data Type

The following Gantt Diagram, Load Diagram and Day planning Diagram are three examples inherited from this Graphic Abstract Data Type.

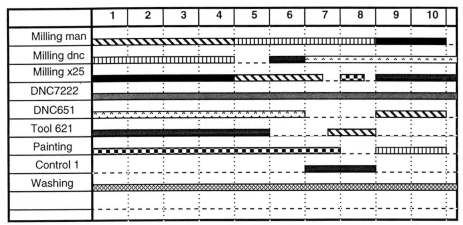

	1	2	3	4	5	6	7	8	9	10
Milling man										
Milling dnc										
Milling x25										
DNC7222										
DNC651										
Tool 621										
Painting										
Control 1										
Washing										

Figure 41 : Example of Gantt Diagram

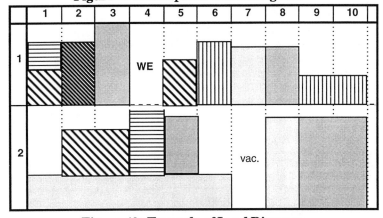

Figure 42 : Example of Load Diagram

	Monday	Tuesday	Wednesday	Thursday	Friday
1	8h-15h	10h30 -20h	8h-10h / 13h30 -19h	8h-20h	Day of rest

Figure 43 : Example of Day planning Diagram

This generic approach presents a number of advantages:

- it gives added precision to the model,
- it makes generalisations easier,
- it helps software development,
- it defines a software component independently of computer languages,
- these software components may be specified by users,
- it facilitates automatic generation of programs.

3.6. Elaboration of a suitable programming language.

The choice of a suitable programming language has to be made according to the following criteria :

- **Modelling capability** : the programming data types of the language should facilitate the description of the various models used in the Software; to program lists and trees, the data type pointers and handles must exist.
- **Rapidity** : to compute the scheduling algorithm or to search in graphs, the language have to be very quick. So, it is not possible to use an Artificial Intelligence Programming Environment based on Lisp or Prolog. We therefore need a procedural language.
- **Objects** : the conceptual models for Controlled FMS, for Software Architecture or for Interactive Diagrams are based on objects. Therefore, we need an object-oriented language to program these models.
- **Reusability of tools** : This criterion may be regarded as reusability of software components :
  - within one same software,
  - from an FMS Planning and control software to another one,
  - from an FMS Planning and control software to another Planning and control software applied to a correlated problem such as Course Scheduling, Project planning,...
  For this, the language must support the notion of encapsulation to build software components (with one interface part and one implementation part).
- **Portability** : the software must be portable from a platform (hardware, operating system, multi-windowing environment) to another one.
- **Interactivity** : this criterion is very much linked to the multi-windowing environment. Today, this criterion is often incompatible with portability; choosing a common object library is a way to increase portability.

C++, ADA or an Object Pascal may be good choices for the basic language. The choice of the Basic Object library will be made so that it runs on several computers and operating systems.

3.7. Development methodology.

Once a processing module or program is specified, we need a method to write the program, to modify it and to rationalise its structure in order to make it reusable [SAR.93].

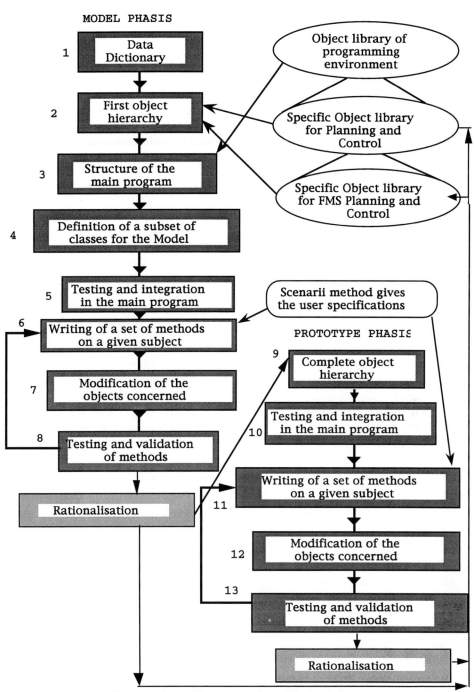

Figure 44 : Programming Methodology

Today, we are not able to design a final interactive program directly. Even with methods like the Scenarii Method, it is necessary to test the feasibility of the most interactive functions with the future users.

So, the development methodology is an iterative process consisting of two main phases [BER 91]:

> - the Model phasis aims at validating the main concepts of the program, to prove its feasibility. This phasis consists in defining a first hierarchy of object classes and testing its capability to model simple cases. Only the main interactive functionality is programmed to show the principle of Man-Machine Interactions.
> - the Prototype phasis gives the first program that can be used for real cases. The Prototype must realise all the functionalities described by the scenarii method.

The flow diagram shows the successive steps of the method and the main supports of these phases:

> - the Scenarii Method gives the user specifications,
> - three object libraries are used to build the program:
>> - Basic Object library of programming environment,
>> - Specific Object library for Planning and Control,
>> - Specific Object library for FMS Planning and Control.

The Basic Object library belongs to the programming environment and must not be modified by the program. This library allows to manage the basic elements of the software: files, mouse and keyboard, menus, screens and windows, basic controls of the windows such as buttons, scroll bar, pictures,... All the programs in the FMS Software must use the same Basic Object library.

The Specific Object library for Planning and Control concerns objects such as Interactive Gantt Diagrams, Calendars, Resources, Products, Routes, Tasks,... Many concepts of the Kernel Model are inherited from objects in this library. For the purpose of FMS Control they can be modified by inheritance to fit better with FMS modelling.

The Specific Object library for FMS Planning and Control contains all the objects designed specifically for our software. They may be inherited from the two other libraries. The objects must be able to be used in several programs of the software.

The use of this three related libraries is a good manner to save time (and money) when programming an FMS Planning and Control Software. Reusability of objects saves time, in addition, encapsulation of code avoids wasting time in debugging programs.

MODEL PHASIS

Step 1 :   Data Dictionary

> It is a classical step in program analysis. It consists in describing all the data entities with their attributes and relations. Most of these entities have to become objects.

Step 2 :   **First object hierarchy**

The First object hierarchy is designed from the Data Dictionary. It is an object hierarchy specific to our problem. The rules to choose which objects have to be created are issued from the two specific Object libraries and from the know-how of object programming in the environment that has been chosen.

Step 3 :   **Structure of the main program**

The Structure of the main program is directly related to the programming environment and to its Basic object library. Each environment has its own way to organise the skeleton of a main program. This step is also concerned with the design of main menus and windows.

Step 4 :   **Definition of a subset of classes for the Model**

The purpose of the Model is to test the feasibility of the main concepts concerning the program. So, this step selects a set of classes that are significant for these concepts.

Step 5 :   **Testing and Integration in the main program**

The aim is to achieve compatibility between the object hierarchy designed for our FMS Control problem and the more standard object structure of the main program.

Step 6 :   **Writing of a set of methods on a given subject**

The global structure of the program is now defined. The programming job begins. Steps 6, 7 and 8 are a loop. For each loop, a subject is chosen (for instance, saving objects in the Database) and all the related methods are programmed for all the objects concerned.

Step 7 :   **Modification of the objects concerned**

Programming an object method may lead to modify an attribute of the object or the interface (external view) of another object.

Step 8 :   **Testing and validation of methods**

This step builds simple cases to use and validate the method algorithms.

PROTOTYPE PHASIS

Step 9 :   **Complete object hierarchy**

From the first object hierarchy and the results of the Model this step builds the object hierarchy specific to FMS. Some attributes are added to model FMS real cases better.

Step 10 :   **Testing and Integration in the main program**

Idem step 5

Step 11:   **Writing of a set of methods on a given subject**

Idem step 6

Step 12 :   **Modification of the objects concerned**

Idem step 7

Step 13 :   **Testing and validation of methods**

Idem step 8

Rationalisation

The rationalisation step aims at creating or updating objects after a development phasis. It is a team work between programmers building different parts of the same software. Even if this step is not included in the development methodology, it is very important to improve both specific libraries. The more extended these libraries will be, the less expensive the programs will be to develop.

4.   Conclusion

FMS Planning and Control is an interdisciplinary field. Required tools come from Operational Research, Control Theory, Software Engineering, Manufacturing Systems Modelling.

To design an efficient Software Architecture we think that it is necessary to have a global and integrated approach of the problem. We therefore propose to view this problem as a Control Problem, as the main objective is to guide the FMS towards manufacturing objectives defined by the upper Decision Levels.

In order to master the complexity of FMS our basic option is to design a convivial Control Software. The second option is to design an Open and Evolutive System.

The programming option to choose an object environment allows to reuse software components and to realise a true Open and Evolutive Software. So, it is a good way to attempt to reduce prohibitive development and maintenance costs.

To continue and to improve the efficiency of the Control System we use a **multiple-agent approach**. To put it shortly, this approach transforms each object into an agent by giving it an objective to reach and its own knowledge. The scheduling algorithm is based on a cooperative agent net. It may be activated according to several strategies depending on the agent summoned. For instance, it is possible to activate only one operation agent or one machine agent. In that case, the agent activated controls the algorithm by exchanging messages with others. So, the scheduling algorithm possesses several strategies according to the agent activated. Each agent is potentially an input point of the algorithm. This operation structure should improve the scheduling algorithm flexibility. On-line and dynamic control will be able to use it to make partial plans. This should also help users to enter theirs decisions more quickly and to obtain partial plans more easily.

5. Bibliography

[ARC.87] ARCHIMEDE B., Développement du noyau de l'outil de simulation SIMGRAI et des outils logiciels pour le traitement des réseaux de données, Mémoire de DEA, Université de Bordeaux 1, Septembre 1987.

[ARC.91-1] B. ARCHIMEDE - Conception d'une architecture réactive distribuée et hiérarchisée pour le pilotage des systèmes de production - Thèse de doctorat - Université de Bordeaux I.1991.

[ARC.91-2] B. ARCHIMEDE, L. PUN, C. BERARD, G. DOUMEINGTS - Real-Time Dynamical Diagnosis in the Control of Flexible Manufacturing System - IMACS Symposium - 7,10 Mai 1991 - Casablanca, Maroc.

[ARC.91-3] L. PUN, B. ARCHIMEDE, C. BERARD, G. DOUMEINGTS - Intelligent Learning-aid for an Intelligent FMS-Monitoring Process - CIRP Workshop on intelligent Learning Manufacturing System - March 6,8 1991 - Budapest.

[BER.81] Ch. BERARD, D. BREUIL, G. DOUMEINGTS & L. PUN. Decision aid in job shop production. 6th International Conference on Production Research, Novisad, Yougoslavie (August 1981).

[BER.83] BERARD Christian, Contribution à la conception de structures logicielles pour le pilotage d'atelier, Thèse de Docteur-Ingénieur, Université de Bordeaux1, Janvier 1983.

[BER 85] C. BERARD, A. GUDEFIN, A. BOURELY, G. DOUMEINGTS. "SIMGRAI : a new tool to simulate manufacturing systems". IMACS, Oslo, Norvège (August 1985).

[BER 87] CHRISTIAN BÉRARD, GUY DOUMEINGTS. - Man-Machine Communication Analysis in a Decision Aid System for Controlling Manufacturing Activities - New Techniques and Ergonomics, Hermes, Paris, 1987, p.68-81.

[BER 89] BERARD Christian, L.T.N. SEGUIN, A. Dupeux, Manuel d'utilisation de SIPA+, laboratoire GRAI, Université de Bordeaux1, Janvier 1991.

[BER 90] C. BERARD, Ph. NGUYEN. Intelligence Artificielle et Productique, IFIP: "Méthode des scenarii: Analyse et Conception des interfaces Homme/Machine", Bordeaux, (Mai 1990).

[BER 91] BERARD Christian, Les Réseaux de Données, rapport interne, laboratoire GRAI, Université de Bordeaux1, Janvier 1991.

[BOC.91] BONNEVAL A., COURVOISIER M., COMBACAU M., FMS Real-time Monitoring: Decision-Making Aspects in Automatic Recovery, Computer Applications in Production and Engineering., IFIP,1991, p. 409/ 416.

[BON. 91] C. BONNEMOY et S.B. HAMMA, La méthode du recuit simulé optimisation globale dans R^n, RAIRO Vol 25, N°5, 1991 (477-496)

[BOW.91]   R.BOWDEN, J.BROWNE, Approach to the production environment design task within factory coordination, Computer Integrated Manufacturing Systems, Vol.4 N°1 February 1991 pp 42-50.

[BOY. 92]   BOY G., Méthodologies et outils pour l'Interaction Homme-Machine, Mémoire d'habilitation, Université Paris 6, Juin 1992

[BUR.86]   BURBIDGE J. L., Production flow analysis, International Journal of Production Research, Octobre 1986.

[COU.89]   COURVOISIER M., VALETTE R., SAHRAOUI A., COMBACAU M., Specification and implementation techniques for multilevel control and monitoring of FMS, Computer Applications in Production and Engineering., IFIP,1989, p. 509/ 516.

[DOU.84]   DOUMEINGTS G., Méthode GRAI: méthode de conception des systèmes en productique, Thèse d'état, Automatique, Université de Bordeaux I, 13 Novembre 84, 519 p..

[DOU.90]   DOUMEINGTS G., Méthode pour concevoir et spécifier les systèmes de production, Laboratoire GRAI, Université de Bordeaux I, CIM 90,Teknea, p. 89 /103.

[DOU.87].   DOUMEINGTS G., DARRICAU D., POUMEYROL E. et ROBOAM M., La méthode GRAI, Cours de l'Université de Bordeaux I, 1986 / 87, 43 p.

[ELO. 92]   A. ELOMRI, Méthodes d'optimisation dans un contexte productique, Thèse de Doctorat de l'Université Bordeaux1, 1992

[GRA.91]   PETER O'GRADY, RAMADURAI SESHADRI, X-Cell intelligent cell Control using object-oriented programming (part I), Computer Integrated Manufacturing Systems, Vol.4 N°3 August 1991 pp 157-163.

[GUY 91]   GUYOT Jean Edouard, Les outils de réaction du modèle PCS (Planification, Conduite, Suivi), laboratoire GRAI, Université de Bordeaux1, Mémoire DEA, Septembre 91.

[HUT.91]   HUTCHISON Jim, LEONG Keong, SNYDER David, WARD Peter, Scheduling approaches for random job shop flexible manufacturing systems, Prod. Res., vol 29,n° 5,1991, p. 1053/ 1067.

[JON.85]   JONES A., MCLEAN C., A Production Control Model for the AMRF, Proceedings of the International ASME Conference on Computers in Engineering

[JON.90]   JONES Albert, SALEH Abdol, A multi-level/ multi-layer architecture for intelligent shopfloor control, Computer Integrated Manufacturing, vol 3 ,n° 1,1990, p. 60/ 70.

[KAL.85]   KALLEL G., PELLET X., BINDER Z., Conduite décentralisée coordonnée d'atelier, R.A.I.R.O., APII, vol 19,n° 4, 1985, p. 371/ 387.

[KAL.85]   KALLEL Ghazali, Proposition d'une COnduite DEcentralisée COordonnée(CODECO) pour un atelier de fabrication, Thèse de Docteur, Institut National Polytechnique de Grenoble, Juin 1985.

[KER.93]   L.KERMAD, C.AUSFELDER, J.P. BOUREY, E CASTELAIN, Integrative approach for a functional specification of FMS Control, Computer Integrated Manufacturing Systems, Vol.6 N°4 November 1993 pp 219-227.

[LEC 90]   LECLERC Christophe, Aide au pilotage: Réalisation d'une architecture multitâches pour le pilotage temps réel de l'atelier, laboratoire GRAI, Université de Bordeaux1, Mémoire d'ingénieur, Septembre 90.

[LEG.89] LE GAL A., Un système interactif d'aide à la décision pour le l'ordonnancement et le pilotage en temps réel d'atelier, Thèse de Doctorat, Université Paul Sabatier de Toulouse, Septembre 1989.

[MAD.90] H.K.MADHUSUDHANA, KRIPA SHANKA, FMS simulation model using PC-SIMSCRIPT II.5, Computer Integrated Manufacturing Systems, Vol.3 N°3 August 1990 pp 150-156.

[MAH.90] MAHMOODI F., DOOLEY K. J., STARR P., An investigation of dynamic group scheduling heuristics in a job shop manufacturing cell, Prod. Res., vol 28,n° 9,1990, p. 1695/ 1711.

[MER.93] K. MERTINS, W.SÜSSENGUTH, Integrated information modelling for CIM, Computer Integrated Manufacturing Systems, Vol.4 N°3 August 1991 pp 123-131.

[MIL 87] MILIN : "Amélioration de solutions d'ordonnancement pour l'aide au pilotage d'ateliers", Thèse d'université: Automatique: Bordeaux I:1987.

[MON.90] MONTAZERI M., VAN WASSENHOVE L. N., Analysis of scheduling rules for an FMS, Prod. Res., vol 28,n° 4,1990, p. 785/ 802.

[MUK.91] MUKHOPADHYAY S. K., MAITI B., GARG S., Heuristic solution to the scheduling problems in flexible manufacturing system, Prod. Res., vol 29,n° 10,1991, p. 2003/ 2024.

[NEE.90] Francis NEELAMKAVIL, Development of graphical software for CIM applications, Computer Integrated Manufacturing Systems, Vol.3 N°3 August 1990 pp 157-162.

[PAC 92] PACHET F., Représentation de connaissances par objets et règles: le système NéOpus, Thèse de Doctorat : Informatique option Intelligence Artificielle, Université Paris 6, Septembre 1992

[PEL.85] PELLET X., Sur la hiérarchisation des décisions. Application à la conduite d'atelier, Thèse de Docteur, Institut National Polytechnique de Grenoble, 1985.

[PLO.84] PLOBNER Alain, Méthodes de conception et de développement des logiciels de pilotage d'unités de production, Thèse de Docteur ingénieur, Université de Bordeaux 1, 1984.

[POR.88] PORTMANN M. C., Méthodes de décomposition spatiale et temporelle en ordonnancement de la production, R.A.I.R.O., APII, vol 22,n° 5, 1988, p. 439/ 451.

[POR.87] PORTMANN M.C.,Méthodes de décomposition spatiale et temporelle en ordonnancement de la production, Thèse de doctorat d'état ès sciences, Université de Nancy 1, Septembre 1987.

[PUN 83] L. PUN, G. DOUMEINGTS, A. BOURELY. "GRAI Methodological Approach to the design of Flexible Manufacturing". ICPR 83, Windsor, Canada (August 22-24 1983).

[PUN 84] L. PUN. Systèmes industriels d'Intelligence Artificielle - Outils de Productique. Editions : Editest en français (1984), Plenium en anglais (1986), Friendship en chinois (1986).

[ROU.88] ROUBELLAT F., THOMAS V., Une méthode et un logiciel pour l'ordonnancement en temps réel d'ateliers, R.A.I.R.O., APII, vol 22,n° 5, 1988, p. 419/ 438.

252

[SAN.91] S.P.SANOFF, D.POILEVEY, Integrated information processing for production scheduling and control, Computer Integrated Manufacturing Systems, Vol.4 N°3 August 1991 pp 157-163.

[SAR.93] P.SARGENT, Inherently flexible cell communications : a review, Computer Integrated Manufacturing Systems, Vol.6 N°4 November 1993 pp 244-259.

[SUN.89] SUN QI-ZHI, Expert simulation for online production control, Computer Integrated Manufacturing Systems, Vol.2 N°3 August 1989 pp 172-180.

[THO.80] THOMAS V., Aide à la décision pour l'ordonnancement en temps réel d'atelier, Thèse de 3° cycle, Université Paul Sabatier, Toulouse, 1980.

[THO.90] THOMESSE Jean-Pierre, Les services de terrain FIP et l'intégration dans les systèmes automatisés, Colloque International CIM90, 12-14 juin 1990, p. 565/ 573, Bordeaux, France.

[TIT.79] TITLI (A.).- Analyse et commande des systèmes complexes. - Toulouse, Cepadues Editions, 1979, 237 p.

[VEE.93] D.VEERMANI, B.BHARGAVA, M M BARASH, Information system architecture for heterarchical control of large FMSs, Computer Integrated Manufacturing Systems, Vol.6 N°2 May 1993 pp 76-92.

[WAR.91] H.J.WARNECKE, W.DANGELMAIER, Functions and interfaces in a flexible production planning and control system, Computer Integrated Manufacturing Systems, Vol.4 N°1 February 1991 pp 31-41.

Flexible Manufacturing Systems: Recent Developments
A. Raouf and M. Ben-Daya (Editors)
© 1995 Elsevier Science B.V. All rights reserved.

Control System Design for Flexible Manufacturing Systems[1]

F. L. Lewis[a], H.-H. Huang[a], O. Pastravanu[b], and A. Gürel[c]

[a]Automation and Robotics Research Institute, The University of Texas at Arlington, 7300 Jack Newell Blvd. S, Ft. Worth, Texas 76118
[b]Dept. Automatic Control and Ind. Inf., Polytechnic Institute of Iasi, Str. Horia 7-9, 6600 Iasi, Romania
[c]Electrical Eng. Dept., Eastern Mediterranean Univ., Gazi Magosa, Cyprus

A modern systems theory point of view is offered for the design of sequencing controllers for flexible manufacturing systems, whereby the controller is considered as separate from the workcell. An algorithm is given to design an intelligent rule-based controller with inner loops where no shared-resource conflict resolution is required, and outer loops that require extra control inputs to resolve conflict where needed. The controller design is based on standard manufacturing tools such as the BOM, task sequencing matrix, and resource requirements matrix; it is described by matrix equations over a nonstandard algebra. The equations of the closed-loop manufacturing system can be used to derive the Petri net or max/plus dioid description of the system for rigorous performance analysis.

1   Introduction and Background

Although a great deal has been done on the analysis of discrete event dynamic systems (DEDS), very little has been done in the way of design for guaranteed performance. Popular current approaches for analysis of cellular and flexible manufacturing systems (FMS) include probability models, Petri nets (PN) [7], [19], [30], [32], [36], [40], [41], [43], [47], language-based approaches [37], sample path approaches [12], [13], object-oriented techniques [9], rule-based expert systems and multi-agent approaches [4], and constrained search techniques [42]. See [22] for a survey. An approach that casts DEDS into the framework of a linear system over the max/plus or dioid algebra is given in [6].

Unfortunately, it is very difficult to write down, for instance, the PN description of any real manufacturing system. Various authors are making efforts to correct such problems, but they generally need to define hierarchical PN, colored PN, extended PN, and other refined notions [10], [17], [31], [38], [46].

---

[1]The research is supported by NSF grants MSS-9114009 and IRI-9216545.

On the other hand, in manufacturing engineering there are tools available for manufacturing scheduling such as the partial assembly tree [45], bill of material (BOM), task sequencing matrix [8], [44], and resource requirements matrix [21]. The relation of these to DEDS approaches using PN and max/plus is obscure. A body of work exists for shared-resource conflict resolution in industrial engineering, namely, the work on dispatching and scheduling (e.g. [21], [33]). Standard dispatching rules show how to operate manufacturing cells in the presence of limited resources such as pallets, transport robots, machines. Recent work [20], [27], [28] brings a system theory flavor into manufacturing dispatching, with performance often being guaranteed via mathematical proofs. Unfortunately, it is not fully understood how such dispatching results fit into the design paradigm of PN and other DEDS analysis approaches.

We confront DEDS controller design from a rigorous point of view by examining FMS from a systems theory perspective [3]. Thus, a distinction is made between the plant or workcell, comprised of the machines and resources to be controlled, and the controller, consisting of the decision-making scheduler. Briefly, all the places in a PN are part of the plant, and all the transitions are part of the controller. A design algorithm for discrete event (DE) control systems is given that affords a very convenient approach to the initial design of FMS, as well as redesign to incorporate modifications (e.g. resource reallocation, failure modes).

The proposed controller has an intelligent rule-based structure that is explicitly given in terms of the problem structure and standard industrial engineering tools. It is based on the task sequencing matrix and the resource requirements matrix. It consists of inner loops, where no resolution of shared-resource conflicts is needed, and outer loops with additional control inputs to resolve conflict. In the outer loops, all the standard dispatching rules can be used. The FMS controller is described in terms of matrix equations over a nonstandard algebra, and affords a framework for rigorously checking the effectiveness of the dispatching rules proposed in terms of deadlock removal and effective resource allocation.

Given the FMS controller, it is easy to determine the corresponding PN, so that PN techniques can be used to analyse the closed-loop system to verify suitable performance in terms of absence of deadlock, and so on. The controller equations can also be manipulated to yield the max/plus (dioid) description of the closed-loop FMS. The result is a step-by-step procedure for constructing a PN for real workcells, including their controllers, starting with the standard manufacturing engineering tools of the BOM, task sequencing matrix, and resource requirements matrix.

We begin by discussing some basic building blocks for DE control systems.

2 Discrete Event Controller Design Components

To provide systematic techniques for the design of discrete event (DE) controllers, it is important to have some basic modular functional design components. In this section we provide such components, representing them in terms of Petri net transitions and places, and also in terms of sequencing matrices.

The sequencing matrix has been used for specifying sequencing requirements in task planning [8], [44]. In this matrix, the columns and rows correspond to tasks, and entries of '1' in a row indicate the tasks or conditions that are prerequisites for that task. Thus, an (i,j) entry of 1 indicates that task j is a prerequisite for task i. For instance, the matrix in Fig. 2a shows that to start task c it is first necessary to complete both tasks a and b.

We use the sequencing matrix in a more general and rigorous way in this chapter, allowing two sorts of operations. Fig. 2 uses the standard 'and' operation (denoted $\wedge$), while Fig. 3 uses the 'or' operation (denoted $\vee$). It is necessary to specify the base operation associated with any given sequencing matrix.

In DE controllers it is useful to speak in terms of task paths and resource loops. The motivation is that tasks are performed on parts which enter and leave the cell, while resources are assigned and subsequently released on task completion, being then available for reassignment at the same location.

2.1 Discrete Event System Design Elements

We begin with the most basic DE design elements, which can be combined to form more complex components.

2.1.1 Series Sequencing Element

A standard sequencing element for series tasks is given in Fig. 1. This element is useful in the task path for describing temporal precedence requirements among tasks.

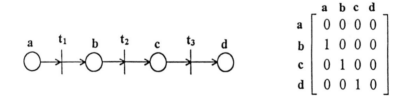

Figure 1: Series sequencing element

2.1.2 Logical Design Elements

There are four logical design elements, as shown in Fig. 2 and Fig. 3. The case of three places is shown; they can all be generalized to more places.

The two 'and' elements In Fig. 2 are useful in the task path. The logical element

$$If\ (a.and.b)\ then\ (c) \tag{2.1}$$

will be called a conditional transition and is useful for synchronizing events prior to commencement of a task (e.g. part present and resource available). The element

$$If\ (a)\ then\ (b.and.c) \tag{2.2}$$

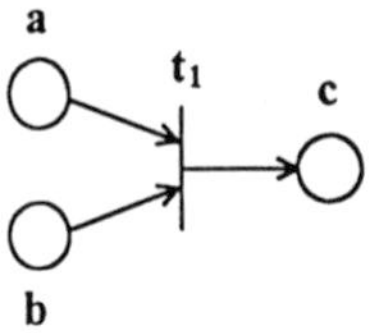

(Base operation ∧)

(a) IF (a .and. b) THEN (c)

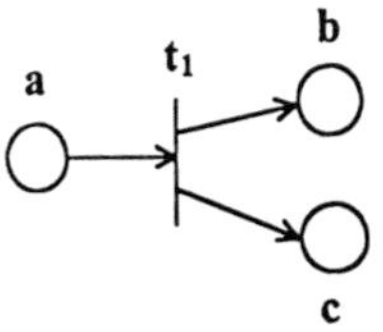

(Base operation ∧)

(b) IF (a) THEN (b .and. c)

Figure 2: 'AND' design elements

is important at the end of a task, where several actions are next required (e.g. request next task and release resource).

As a word on terminology, the left-hand sides of rules are called the antecedent (or conditions) and the right-hand sides are called the consequent (or actions).

The two 'or' elements in Fig. 3 are generally useful in resource loops when there are shared resources (i.e. a robot arm that must perform two tasks). The element

$$If \ (a) \ then \ (b.or.c) \qquad (2.3)$$

is useful in assigning shared resources (e.g. if robot carrier is available then perform transport task b or transport task c). This element is generally ill-defined, involving conflicting requirements for the same resource, and requires additional decision-making to remove the conflict. This involves conflict analysis and conflict resolution.

The element

$$If \ (a.or.b) \ then \ (c) \qquad (2.4)$$

is important in the resource loops when releasing shared resources (e.g. If task a is complete or task b is complete, then set robot carrier idle).

2.1.3   Duality

Design elements (2.1) and (2.3) are said to be dual, where duality is defined by interchanging 'and' with 'or' and reversing the rule conditionality (i.e. swap antecedent and consequent). Note that their sequencing matrices are transposes of each other, though under different base operations.

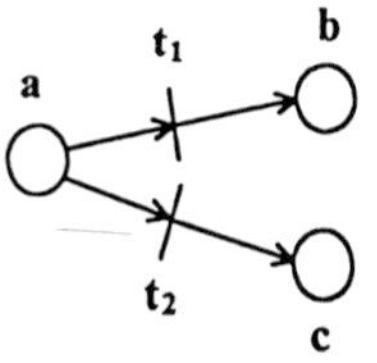

$$\begin{array}{c} \begin{array}{ccc} a & b & c \end{array} \\ \begin{array}{c} a \\ b \\ c \end{array} \left[\begin{array}{ccc} 0 & 0 & 0 \\ 1 & 0 & 0 \\ 1 & 0 & 0 \end{array} \right] \end{array}$$

(Base operation $\vee$)

(a) IF (a) THEN (b .or. c)

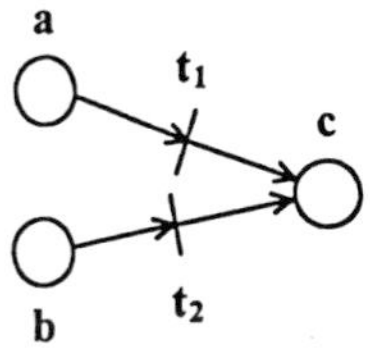

$$\begin{array}{c} \begin{array}{ccc} a & b & c \end{array} \\ \begin{array}{c} a \\ b \\ c \end{array} \left[\begin{array}{ccc} 0 & 0 & 0 \\ 0 & 0 & 0 \\ 1 & 1 & 0 \end{array} \right] \end{array}$$

(Base operation $\vee$)

(b) IF (a .or. b) THEN (c)

Figure 3: 'OR' design elements

A similar duality holds for (2.2) and (2.4). The duality of these elements results in the fact that part routing and job scheduling are duals as well; such notions are beginning to appear in the literature [39].

2.2 Combining Logical Design Elements

The logical design elements can be combined in several ways into basic design components or blocks.

2.2.1 Task Performance and Resource Scheduling

In the task path, a useful block is shown in Fig. 4, which shows parallel operations. Another useful block is the task synchronization in Fig. 5: For instance, to perform task c, task a must be complete and resource b must be available. When task c is done, task d is requested and resource e is released. (If resources b and e are the same, then a resource loop should be closed by identifying b and e as the same place.)

In the resource loop, the useful design block is shown in Fig. 6, which depicts a shared resource: For instance, when task a or task b is done, release resource c. Then, assign the resource either to task d or to task e. Unfortunately, this block causes most of the thorny problems in the representation and analysis of DEDS. In fact, most references disallow this case [6], [19]. A Petri net with no 'or' elements such as those in Fig. 3 is called decision-free and is nothing but an event graph. These admit simplified analysis techniques. Equally unfortunately, the 'or' elements are often required to model practical

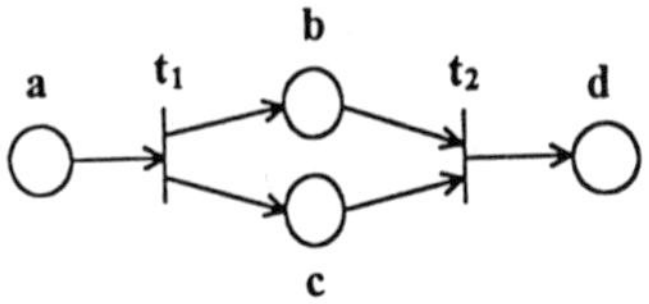

$$\begin{array}{c} & \begin{array}{cccc} \mathbf{a} & \mathbf{b} & \mathbf{c} & \mathbf{d} \end{array} \\ \begin{array}{c} \mathbf{a} \\ \mathbf{b} \\ \mathbf{c} \\ \mathbf{d} \end{array} \left[\begin{array}{cccc} 0 & 0 & 0 & 0 \\ 1 & 0 & 0 & 0 \\ 1 & 0 & 0 & 0 \\ 0 & 1 & 1 & 0 \end{array} \right] \end{array}$$

(Base operation $\wedge$)

Figure 4: Parallel sequencing block

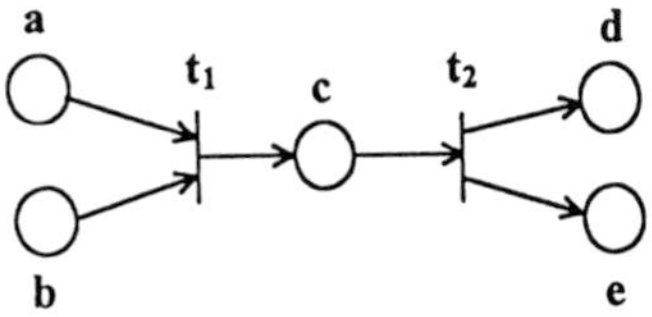

$$\begin{array}{c} & \begin{array}{ccccc} \mathbf{a} & \mathbf{b} & \mathbf{c} & \mathbf{d} & \mathbf{e} \end{array} \\ \begin{array}{c} \mathbf{a} \\ \mathbf{b} \\ \mathbf{c} \\ \mathbf{d} \\ \mathbf{e} \end{array} \left[\begin{array}{ccccc} 0 & 0 & 0 & 0 & 0 \\ 0 & 0 & 0 & 0 & 0 \\ 1 & 1 & 0 & 0 & 0 \\ 0 & 0 & 1 & 0 & 0 \\ 0 & 0 & 1 & 0 & 0 \end{array} \right] \end{array}$$

(Base operation $\wedge$)

Figure 5: Synchronization for task block

FMS where scarce or expensive resources (e.g. robot arm) may be shared for several tasks.

It is generally important to use only the canonical elements shown in Fig. 2 and Fig. 3, as the next example on deadlock shows.

2.2.2 Deadlock

Deadlock can occur in a FMS with shared resources; it is an undesirable condition where there are parts in the workcell but no jobs can be performed. For instance, a transport robot R1 is required to unload a machine M1, but R1 is holding a part that it cannot release until machine M1 is unloaded. Deadlock often requires manual intervention to restore the FMS to operation. If a system has potential deadlock, then incorrectly servicing requests for shared resources (e.g. incorrect dispatching) can make it occur. That is, improper management of shared resource conflicts can manifest deadlock. There are various techniques for making it impossible for a system to deadlock, that is, for making a system *deadlock-free*. It is advisable to take steps to avoid deadlock prior to implementing any dispatching rules. The controller design algorithm in Section 4 does this when it is possible. Then, though there are still shared-resource conflicts, their resolution will never deadlock the system.

Example 2.1: Avoiding Deadlock

Fig. 7 shows a FMS with two operations, O1 performed by a CNC machine tool and O2 performed by drilling robot DR. Parts arrive at PI and depart at PO. Two moves are needed, M1 to carry the parts to the CNC tool, and M2 to carry them to the drilling robot. The label 'I' means a resource is idle.

Assigning two transport robots R1 and R2, one for each move, as shown in Fig. 7

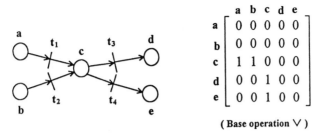

$$\begin{array}{c} \\ a \\ b \\ c \\ d \\ e \end{array} \begin{array}{ccccc} a & b & c & d & e \\ \left[\begin{array}{ccccc} 0 & 0 & 0 & 0 & 0 \\ 0 & 0 & 0 & 0 & 0 \\ 1 & 1 & 0 & 0 & 0 \\ 0 & 0 & 1 & 0 & 0 \\ 0 & 0 & 1 & 0 & 0 \end{array}\right] \end{array}$$

(Base operation $\vee$)

Figure 6: Shared resource block

results in a suitable strategy with no deadlock. Note that in this figure there appear transitions of a form not shown in Figs. 2, 3.

Unfortunately, transport robots are expensive. Using the same rule structure as in Fig. 7 and assigning only one transport robot R for both moves yields Fig. 8. This figure gives deadlock if there are tokens in places M1 and O1. The meaning is that a part is being held by the transport robot to load into the CNC tool in preparation for O1, but there is also a part in the CNC tool waiting to be removed by the transport robot. This reveals a problem unless the transport robot is ambidextrous.

Traditional approaches to avoiding deadlock [46] would require the initialization of R1 with two tokens. This puts us back to two transport robots. We call this a shared resource pool. An alternative is shown in Fig. 9.

This figure shows how to avoid deadlock without increasing the number of resources by *modifying the antecedents of the synchronization rules.* Thus, the availability of the CNC tool is checked at transition t1, before move M1 is initiated. There is now no deadlock using only one transport robot. Note that the modified DE controller structure has only elements like those in Figs. 2, 3.

This deadlock-free structure is achieved by moving the antecedent conditions on idleness of the nonshared resources upstream (e.g. move condition CNC= I from t2 to t1) so that *the transitions releasing the shared resources are never conditional.* In fact, this corresponds exactly to *kanban feedback,* since a pull, corresponding to the availability of the CNC tool, is required to request part entry to the cell. A rigorous analysis of this can be carried out [11], [34].

Note that, though the system is now deadlock-free, shared resource conflict can still occur, so that some sort of dispatching rule is still needed to assign the transport robot to M1 or M2 when both moves are requested simultaneously. ■

2.2.3 Buffers and Queues

An important design block is the buffer or queue shown in Fig. 10. There, the maximum queue length L corresponds to the number of initial tokens in place d. These initial tokens may be considered as kanban cards which must accompany the parts into the queue (place b). When the cards are expended, no more parts may enter the queue.

The realistic situation occurs when place b has a processing time T_b which must elapse before tokens may fire transition t_2, or when an additional condition (not shown) is needed to fire t_2 besides the presence of a token in place b.

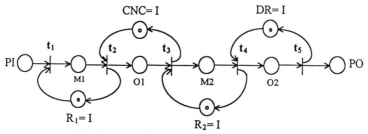

Figure 7: FMS with no shared resources

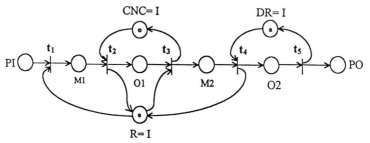

Figure 8: FMS with shared resources and deadlock

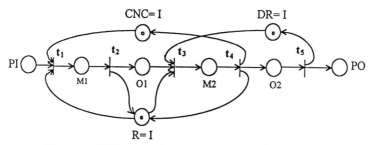

Figure 9: FMS with shared resources and no deadlock

The buffer is nothing but the series sequencing element with an extra row and column for the kanban card place d. The next example shows the usefulness of buffers in decoupling (buffering) series operations.

Example 2.2: Adding a Buffer for Decoupling

Fig. 11 shows a FMS with two series tasks: move M1 (executed by transport robot R) brings the part to a CNC tool which performs an operation O1. In the figure, it is necessary for CNC to be idle prior to executing M1 (see Example 2.1); this could result in a part being held in place PI for some time, and thus in a backlog of requests in the FMS upstream of place PI.

To correct this, one may insert a buffer B as shown in Fig. 12. Now, CNC idle is no longer required for move M1, so that the execution of task one is no longer dependent on the state of task two. A suitable value of the maximum queue length L depends on the

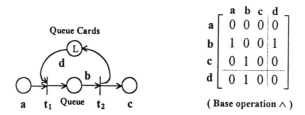

$$\begin{array}{c} \begin{array}{cccc} \text{a} & \text{b} & \text{c} & \text{d} \end{array} \\ \begin{array}{c} \text{a} \\ \text{b} \\ \text{c} \\ \text{d} \end{array} \left[\begin{array}{cc|cc} 0 & 0 & 0 & 0 \\ 1 & 0 & 0 & 1 \\ 0 & 1 & 0 & 0 \\ \hline 0 & 1 & 0 & 0 \end{array} \right] \end{array}$$

(Base operation $\wedge$)

Figure 10: Buffer design block

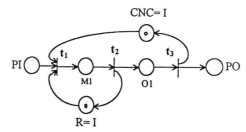

Figure 11: Series FMS

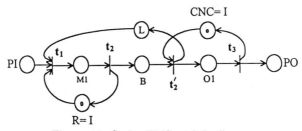

Figure 12: Series FMS with buffer

arrival rate of parts into PI as well as the execution time of O1. In fact, adding a buffer can also often correct deadlock problems if L is large enough.

Note that transition t_2' is not in a standard form shown in Figs. 2.2, 2.3. The buffer appears to be one case where nonstandard elemental forms are acceptable. ∎

3   Representing Discrete Event Systems

At this point it is necessary to discuss three approaches to representing discrete event systems: Petri nets (PN), max/plus algebra, and sequencing matrices/resource requirements matrices. We shall use the manufacturing workcell shown in Fig. 13 as an illustrative example; though simple, modifications of this workcell reveal several problems of manufacturing control, including deadlock, shared resources, and conflict resolution dispatching.

In the workcell, there are 2 machines and 1 buffer; 2 robots unload the 2 machines. The buffer has a finite length of 2, and there are 4 pallets. Machine 1 is loaded with a part (M1P) when a part comes into the cell (PI), pallets are available (PA), and machine 1 is available (M1A). Robot 1 unloads machine 1 (RU1) when the machine operation M1P

262

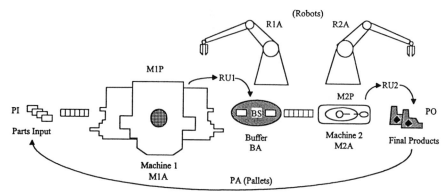

Figure 13: Illustrative FMS workcell

is completed and robot 1 is available (R1A). The part is put into the buffer (BS) when move RU1 is complete and buffer space is available (BA). Machine 2 is loaded with a part (M2P) when a part is in the buffer (BS) and machine 2 is available (M2A). Robot 2 unloads machine 2 (RU2) when the machine operation M2P is completed and robot 2 is available (R2A); then, a product is sent out of the cell (PO) and a pallet is released.

3.1 Petri Nets

It is assumed that the reader has a basic familiarity with Petri nets [30], [36]. However, some PN notions need to be clarified and refined for application to practical FMS. We use timed PN where the places have associated duration times. The transitions are all immediate.

The examples so far presented reinforce the importance of the notions of task path and resource loop. Moreover, an important distinction is evident between task or operation places and resource places (which occur in loops off of the main task path) [15], [46], [47]. It is clear that task places generally need no initial markings, while resource places need initial markings corresponding to the number of resources available for that resource loop. If enough initial tokens are not available in a given loop, deadlock can result if the antecedent structure of the synchronization rules is not correct.

A PN is said to be *conservative* if its total number of tokens is constant. This is indeed not a very useful concept in FMS design. However, the number of tokens in any loop corresponding to a nonshared resource should be constant. Moreover, the sum of tokens in all the loops corresponding to a single shared resource pool should be constant.

Example 3.1: Petri Net Representation of FMS

A Petri net representation of the manufacturing cell of Fig. 13 is shown in Fig. 14. Initial markings are needed on the resource places, denoting initially available resources. There are no shared resources in this example.

It is clear that drawing PN for complex manufacturing systems is not easy. There is no algorithm for PN design, therefore, different design engineers are doomed to produce different PN representations for the same system. Various authors are making efforts to correct such problems, but they generally need to define hierarchical PN, colored PN, extended PN, and other refined notions [10], [17], [31], [38]. The notions of *top-down* and *bottom-up* PN design [15], [46], [47], [7] are very useful. In Section 4 we present a direct

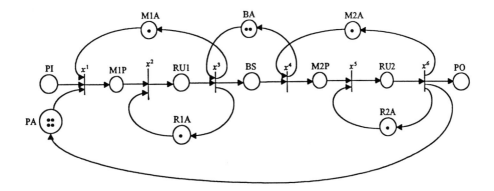

Figure 14: Petri net representation of the workcell

algorithm for FMS controller design from which it is very straightforward to derive a PN representation.

■

3.2 Max/Plus Algebra

The max/plus algebra [6] is a formalism for representing DEDS in the form of a linear system of dynamical equations, which for the case of a delay of one appears as

$$x_{k+1} = Ax_k \oplus Bu_k \qquad (3.5)$$

where k is the iteration index. The state variable x_k is an n-vector with components $[x_k^1 \ x_k^2 \ \cdots \ x_k^n]^T$, where superscript '$T$' denotes transpose. It is straightforward to draw a correspondence between such a system and a PN representation if there are no 'or' elements (i.e. no shared resources and no choices in the task path). The value of x_k^i denotes the time of the k-th firing of transition i. Thus, each transition t_i in an PN has an associated state component x_k^i. The input u_k is an m-vector representing the arrival times of external tokens into the PN.

Matrices A and B contain the duration times of the PN places. The standard matrix multiplication, where A (resp. B) has elements a_{ij} (resp. b_{ij}), is written as

$$x_{k+1}^i = \sum_{j=1}^{n} a_{ij} x_k^j + \sum_{j=1}^{m} b_{ij} u_k^j \qquad (3.6)$$

It uses two operations: multiplication ($\cdot$) is performed between the matrix elements (e.g. a_{ij}) and the vector elements (e.g. x_k^j). Then, addition ($+$) collects all such products into the value for x_{k+1}^i.

To allow PN to be represented in the form (3.5), it is necessary to replace the two standard operations of multiplication and addition respectively by

$$
\begin{aligned}
&\odot \quad \textit{denotes sum.} \\
&\oplus \quad \textit{denotes max value}
\end{aligned}
\qquad (3.7)
$$

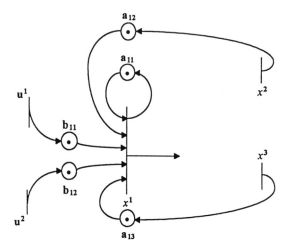

Figure 15: Example of max/plus representation

That is, matrix multiplication is defined over the max/plus algebra. Then, the computations in (3.5) are performed as follows, illustrating one line of (3.5) for the case $n=3$,
$m=2$:

$$x_{k+1}^1 = max(a_{11} + x_k^1, a_{12} + x_k^2, a_{13} + x_k^3, b_{11} + u_k^1, b_{12} + u_k^2), \qquad (3.8)$$

(where $+$ denotes the usual addition). Fig. 15 shows the PN equivalent of this equation.

Suppose x_k^j denotes the times of the k-th firing of transition t_j, and a_{ij} is the duration
time of the place connecting t_j to transition t_i. Then, t_1 can fire for the $(k+1)$-st time
only after t_1, t_2, t_3 have fired for the k-th time, and the associated place duration times
a_{ij} have elapsed. The sums in (3.8) add the transition firing times to the duration times
of their downstream places, and the 'max' operation says all conditions must be satisfied
prior to the firing of transition t_i for the $(k+1)$-st time. (Note, 'max' corresponds to the
'and' operation at t_1).

The issue of initial markings in the max/plus algebra is interesting. Suppose there are
L initial tokens in place a_{12}. Then k tokens can depart a_{12} only after $k - L$ tokens have
arrived. That is, t_1 can fire for the k-th time only after t_2 has fired for the $(k-L)$-th time.
Thus, initial markings correspond to the delays in the equation (3.5). In Fig 15 there is
one initial marking in each place, corresponding to a delay of 1 in each connection. Thus,
t_1 fires for the $(k+1)$-st time only after each t_j has fired for the k-th time.

If transition t_j is not connected to transition t_i, then element a_{ij} is set to $-\infty$, denoted
in max/plus algebra as the max operation zero element ε. This means that the firing of
t_j is not a precondition for the firing of t_i.

It is only possible to write the linear max/plus form (3.5) for a given PN if the PN
has no elements of the form in Fig. 3. That is, there must be a unique transition
downstream of each place (i.e. no forks), and a unique transition upstream of each place
(i.e. no joins). Then, the PN is a marked graph and requires no 'or' operations. An
'or' operation corresponds to the operation 'min', and cannot be accommodated with the

'max' operation to yield a linear system like (3.5). Thus, shared resources are not allowed in the max/plus representation. This is a serious limitation not satisfied by practical FMS; it will shortly be addressed. A partial solution to this problem (which only solves the problems associated with Fig. 3a, i.e. with the forks, not with the joins) is indicated in [5].

Example 3.2: Max/Plus Representation of FMS

For the PN in Fig. 14, the state is defined as

$$x = [x^1 \ x^2 \ x^3 \ x^4 \ x^5 \ x^6]^T,$$

with one component for each transition. Denote the times associated with the places as t with the appropriate subscript. For the machines, these represent operation times (including setup), and for the resources, they represent reset or release times. Using $\odot$ to represent addition and $\oplus$ to represent 'max', the max/plus equations of the transitions are

$$
\begin{aligned}
x_k^1 &= t_{M1A} \odot x_{k-1}^3 & \oplus & \ t_{PA} \odot x_{k-4}^6 & \oplus & \ t_{PI} \odot u_k \\
x_k^2 &= t_{M1P} \odot x_k^1 & \oplus & \ t_{R1A} \odot x_{k-1}^3 & & \\
x_k^3 &= t_{RU1} \odot x_k^2 & \oplus & \ t_{BA} \odot x_{k-2}^4 & & \\
x_k^4 &= t_{BS} \odot x_k^3 & \oplus & \ t_{M2A} \odot x_{k-1}^6 & & \\
x_k^5 &= t_{M2P} \odot x_k^4 & \oplus & \ t_{R2A} \odot x_{k-1}^6 & & \\
x_k^6 &= t_{RU2} \odot x_k^5 & & & &
\end{aligned}
$$

The delays in these equations correspond to the initial markings as noted in Fig. 14. The time of the k-th part entry is denoted by the input variable u_k. The time of the k-th part output is given by the output variable

$$y_k = t_{PO} \odot x_k^6,$$

where t_{PO} is the time needed to get the part out of the cell (e.g. by conveyor).

The max/plus matrix representation is therefore given by

$$
\begin{bmatrix} x_k^1 \\ x_k^2 \\ x_k^3 \\ x_k^4 \\ x_k^5 \\ x_k^6 \end{bmatrix}
=
\begin{bmatrix}
\varepsilon & \varepsilon & \varepsilon & \varepsilon & \varepsilon & \varepsilon \\
t_{M1P} & \varepsilon & \varepsilon & \varepsilon & \varepsilon & \varepsilon \\
\varepsilon & t_{RU1} & \varepsilon & \varepsilon & \varepsilon & \varepsilon \\
\varepsilon & \varepsilon & t_{BS} & \varepsilon & \varepsilon & \varepsilon \\
\varepsilon & \varepsilon & \varepsilon & t_{M2P} & \varepsilon & \varepsilon \\
\varepsilon & \varepsilon & \varepsilon & \varepsilon & t_{RU2} & \varepsilon
\end{bmatrix}
\begin{bmatrix} x_k^1 \\ x_k^2 \\ x_k^3 \\ x_k^4 \\ x_k^5 \\ x_k^6 \end{bmatrix}
$$

$$
+
\begin{bmatrix}
\varepsilon & \varepsilon & t_{M1A} & \varepsilon & \varepsilon & t_{PA} z^{-3} \\
\varepsilon & \varepsilon & t_{R1A} & \varepsilon & \varepsilon & \varepsilon \\
\varepsilon & \varepsilon & \varepsilon & t_{BA} z^{-1} & \varepsilon & \varepsilon \\
\varepsilon & \varepsilon & \varepsilon & \varepsilon & \varepsilon & t_{M2A} \\
\varepsilon & \varepsilon & \varepsilon & \varepsilon & \varepsilon & t_{R2A} \\
\varepsilon & \varepsilon & \varepsilon & \varepsilon & \varepsilon & \varepsilon
\end{bmatrix}
\begin{bmatrix} x_{k-1}^1 \\ x_{k-1}^2 \\ x_{k-1}^3 \\ x_{k-1}^4 \\ x_{k-1}^5 \\ x_{k-1}^6 \end{bmatrix}
+
\begin{bmatrix} t_{PI} \\ \varepsilon \\ \varepsilon \\ \varepsilon \\ \varepsilon \\ \varepsilon \end{bmatrix}
u_k
\qquad (3.9)
$$

where ε represents $-\infty$ and z^{-1} denotes a delay operator, e.g., $z^{-1}x_{k-1}^4 = x_{k-2}^4$. (There is also a corresponding equation for the output.) This may be denoted as

$$x_k = A_0 x_k \oplus A_r x_{k-1} \oplus B u_{k-1}. \qquad (3.10)$$

where it is understood that matrix multiplication is replaced by standard addition. The matrix $A_r(z^{-1})$ is a function of the delay operator z^{-1}. This is equivalently written in terms of constant coefficient matrices as

$$x_k = A_0 x_k \oplus A_1 x_{k-1} \oplus A_2 x_{k-3} \oplus A_3 x_{k-3} \oplus A_4 x_{k-4} \oplus B u_{k-1}. \qquad (3.11)$$

The matrices A_1, A_2, A_3, A_4 are found from A_r taking into account the delay operator z^{-1}, that is, $A_r(z^{-1}) = A_1 + A_2 z^{-1} + A_3 z^{-2} + A_4 z^{-3}$.

There are two major problems with the max/plus representation. First, it only applies for systems with no 'or' elements, which rules out the important case of shared resources, and it is very difficult to obtain if no PN is available. In Example 3.1 we have already mentioned the problems associated with obtaining a PN representation of a manufacturing cell. In Section 4 we shall present an FMS controller design algorithm that gives the max/plus representation directly without the need for first finding a PN.

In this example, matrix A_0 represents the required sequential order of the tasks (which here is in the series form of Fig. 1), while A_r is closely related to the matrix of resource requirements. In Section 4 we shall show that the task sequencing matrix [8], [44] and the resource requirements matrix [21], both standard manufacturing engineering tools, do in fact provide the basis for a design algorithm for FMS controllers from which can be derived both the PN and max/plus representations. ∎

3.3   Sequencing Matrix and Resource Requirements Matrix

Steward's design structure matrix has been used extensively in representing task sequencing for complex manufacturing processes [8], [44]. Here, we call it the *(task) sequencing matrix*; it represents the temporal sequencing relations among the tasks required to produce a finished product.

The *resource requirements matrix*, as used, for instance, by Kusiak [21] is an important industrial engineering tool in workcell design and dispatching. It tabulates the resources needed to perform each job.

It is not obvious from current practice how these manufacturing engineering tools are related to either Petri nets or the max/plus representation. This problem is corrected in Section 4.

3.3.1   Sequencing Matrix

In the sequencing matrix, the columns and rows correspond to tasks, and entries of '1' in a row indicate the tasks or conditions that are prerequisites for that task. Thus, an (i, j) entry of 1 indicates that task j is a prerequisite for task i. In the matrix A_0 of Example 3.2, replacing all times by 1's and all ε by 0's yields the sequencing matrix associated with that FMS.

Using established techniques of task decomposition, one may derive a suitable sequencing matrix for a given FMS. The sequencing matrix may also be derived from a partial assembly tree [45], which comes, e.g, from the BOM. A simple partial assembly tree is given in Fig. 16. The associated sequencing matrix is

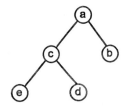

Figure 16: Partial assembly tree

$$F_v = \begin{array}{c} \\ a \\ b \\ c \\ d \\ e \end{array} \begin{array}{c} \begin{array}{ccccc} a & b & c & d & e \end{array} \\ \left[\begin{array}{ccccc} 0 & 1 & 1 & 0 & 0 \\ 0 & 0 & 0 & 0 & 0 \\ 0 & 0 & 0 & 1 & 1 \\ 0 & 0 & 0 & 0 & 0 \\ 0 & 0 & 0 & 0 & 0 \end{array} \right] \end{array}$$

Unfortunately, this sequencing matrix is unsatisfactory, since it does not correspond to an obviously implementable causal time sequence of tasks. Using a reordering of rows and columns described by

$$F_v' = T F_v T^T \qquad (3.12)$$

with T a unitary matrix (i.e. $T^{-1} = T^T$) describing elementary row exchanges, this sequencing matrix F_v can be brought to the *lower triangular form*

$$F_v' = \begin{array}{c} \\ e \\ d \\ c \\ b \\ a \end{array} \begin{array}{c} \begin{array}{ccccc} e & d & c & b & a \end{array} \\ \left[\begin{array}{ccccc} 0 & 0 & 0 & 0 & 0 \\ 0 & 0 & 0 & 0 & 0 \\ 1 & 1 & 0 & 0 & 0 \\ 0 & 0 & 0 & 0 & 0 \\ 0 & 0 & 1 & 1 & 0 \end{array} \right] \end{array} \qquad (3.13)$$

In this modified sequencing matrix, the temporal sequencing of the tasks is evident, for the appropriate causal task sequence is *edcba*. The lower triangular form implies a causal sequence of tasks. Transformations of the form (3.12) are called unitary transformations.

With any sequencing matrix is associated a *sequencing rulebase*. The rulebase for (3.13) is

$$If \quad (d.and.e) \quad then \quad (c)$$
$$If \quad (b.and.c) \quad then \quad (a)$$

The sequencing matrix, plus unitary transformation operations, affords a very convenient method for determining the task sequencing rules, as well as ordering them so they can be executed in a causal fashion. The rulebase is nothing but a set of Petri net transitions.

Extensive discussion of the sequencing matrix is given in [44], where techniques are given to find transformations T that reveal the hidden structure in the sequencing matrix, yielding hierarchical sequencing among the tasks and groups of tasks. Thus, a convenient form results that should give a PN diagram containing subsystems of operations that are connected into the overall system diagram. Unfortunately, the relation between PN, the

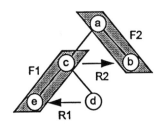

Figure 17: Subassembly tree

max/plus representation, and the sequencing matrix is not yet understood. This issue will shortly be addressed. The connection is important from the standpoint of FMS design, since well-understood techniques exist for deriving the sequencing matrix for a given FMS.

Unfortunately, the sequencing matrix cannot always be made causal. A sequencing matrix that cannot be brought to lower triangular form is

$$
F_v = \begin{array}{c} \\ a \\ b \\ c \\ d \end{array}
\begin{array}{c} a \quad b \quad c \quad d \\
\left[\begin{array}{c|c|c|c}
0 & 0 & 0 & 0 \\ \hline
1 & 0 & 1 & 0 \\ \hline
0 & 1 & 0 & 0 \\ \hline
0 & 0 & 1 & 0
\end{array}\right] \end{array}
$$

This sequencing matrix reveals that tasks b and c must be done in parallel or *simultaneously*. Thus, there is no ordering of the sequencing rulebase that allows causal sequential firing of the rules.

It is always possible to find a transformation T that brings a sequencing matrix to lower *block* triangular form, wherein F_v is lower triangular, but may have blocks on the diagonal instead of single elements. This corresponds to causal interconnections of noncausal (simultaneous) processing subsystems. In manufacturing workcell sequencing, however, the operations can generally be done in a causal or sequential order.

3.3.2 Resource Requirements Matrix

The sequencing matrix does not include any information on the resources needed or available to perform a job; this information is provided by the resource requirements matrix. In Example 3.2, the resource requirements matrix is closely related to matrix A_r.

In Fig. 17 is shown the partial assembly tree of Fig. 16 with fixturing and insertion information as specified by a production engineer and the available resources. It reveals that part d is inserted using tool R1 into part e, which is fixtured in F1, to produce part c. Then, part c is inserted using tool R2 into part b, which is fixtured in F2, to produce part a. The resource requirements matrix corresponding to this situation is

$$F_r = \begin{array}{c} \\ e \\ d \\ c \\ b \\ a \end{array} \begin{array}{cccc} F1 & F2 & R1 & R2 \\ \left[\begin{array}{cccc} 0 & 0 & 0 & 0 \\ 0 & 0 & 0 & 0 \\ 1 & 0 & 1 & 0 \\ 0 & 0 & 0 & 0 \\ 0 & 1 & 0 & 1 \end{array}\right] \end{array} \qquad (3.14)$$

In this situation, there are no shared resources. Shared resources are revealed by the presence of more than one '1' in a single column of F_r; then, dispatching rules [33] must be used when shared-resource conflicts arise. In the case of shared resources, Kusiak [21] has shown that the resource requirements matrix provides the basis for a dispatching algorithm that resolves conflicting requests for the same resource.

Clearly, the sorts of information contained in the sequencing matrix F_v and the resource requirements matrix F_r are not the same. F_v comes directly from a BOM, task decomposition, or partial assembly tree. On the other hand, F_r comes from manufacturing engineering considerations on fixturing, resources available, etc.; some judgement comes into the picture at this point. Moreover, a change in resource allocations effects only F_r, which should aid in redesign or reconfiguration of the FMS. Unfortunately, Petri net and other DE design approaches have been severely limited by not distinguishing between such information of different types; some efforts are now being made to correct this problem (e.g. 'top-down' and 'bottom-up' design [15], [46], [47], [7]). In the next section we provide a FMS controller design scheme that naturally takes such IE design tools into account.

Note that the resource requirements matrix does not include any information on the numbers of resources available (i.e. how many fixtures F1 are there?). On the other hand, this information is included in the max/plus matrix A_r in Example 3.2.

4 An FMS Controller Design Algorithm

In this section we shall give a design algorithm for FMS rule-based controllers. The algorithm is straightforward and *repeatable* so that two engineers will obtain the same results (modulo some design decisions). The algorithm uses the standard manufacturing engineering *task sequencing matrix* and *resource requirements matrix* to produce a rule-based controller for workcell sequencing control. The controller has *inner loops* where no shared-resource conflicts occur, and *outer loops* where dispatching is needed to resolve shared-resource conflicts. The controller is described mathematically in terms of *matrix equations* that implement its decision rules. The controller equations allow a formal approach to design for *deadlock-free operation*.

The Petri net description can be directly determined from the controller matrix equations. Therefore, this algorithm provides a step-by-step method for constructing PN descriptions for manufacturing systems. It contains some formalized aspects of the 'top-down' and 'bottom-up' design techniques in [15], [46], [47]. Once a dispatching rule has been selected for the outer loops, the system is 'decision-free' (c.f. [19], [5]) and an extended max/plus description [6] can be directly determined from the controller matrix equations.

The key issue is the *separation of the task specification and the workcell operations from the sequencing controller*. Thus, the overall task is considered as a partial ordering of jobs (or subtasks), the workcell is considered as a collection of resources available to accomplish those jobs, and the controller is a separate decision-making system that uses information about the workcell status to generate on-line real-time commands for the next jobs to be performed by the workcell. This separation between the plant and the controller is standard in modern systems theory design [3], and the lack of it is the key reason that a convenient standard FMS controller design strategy does not currently exist. This philosophy leads to a real-time feedback controller for workcell scheduling.

4.1 Rule-based Controller for Nonshared Resources Case

The case of nonshared resources does not correspond to practical FMS systems. However, in this case the controller equations are simplified, so that it is suitable for introduction of concepts. The completely general shared-resource case is covered in the next subsection.

In the case of no shared resources a rule-based controller can be designed for a manufacturing system using the following algorithm. The algorithm uses standard tools from industrial engineering to produce a set of equations for the controller that can be computed in real-time on a digital comupter to implement the FMS controller. To cover general FMS, Steward's task sequencing matrix F_v must be complemented by a matrix S_v defined below. Likewise, the resource requirements matrix F_r requires a complementary matrix S_r defined below.

Algorithm 4.1: Controller Design with No Shared Resources

Product Job Sequencing Information:

1. Using the BOM, partial assembly tree, task decomposition techniques, etc, determine the rules needed for task/job sequencing. Resources are not considered here. Define the *controller state vector x* as a vector with one component corresponding to each sequencing rule.

2. Define the *task vector v* as a vector with elements corresponding to all the tasks or jobs involved. Determine the *task sequencing matrix* F_v for the operations. In F_v, all elements are 0, except for 1's included in positions (i, j) if task j is a condition for firing sequencing rule i.

3. Define the *task start matrix* S_v relating v to x. In S_v, all elements are 0, except for 1's included in positions (i, j) if task i is to be started when the conditions for rule j are met. It is often the case in practical FMS that this amounts to setting elements (i, i) equal to 1; that is, task i should start when the conditions for rule i are met.

Raw Material Inputs and Product Outputs:

4. Using information on which input parts are needed for which tasks, define the *part input vector u* as a vector with elements corresponding to all the parts that enter as raw materials into the workcell. Determine the *part input matrix* F_u. In F_u, all elements are 0, except for 1's included in positions (i, j) if entering part j is a needed for task i.

5. Using information on which products result from which tasks, define the *product*

output vector y as a vector with elements corresponding to all the finished products. Determine the *product output matrix S_y*. In S_y, all elements are 0, except for 1's included in positions (i, j) if product i results from task j.

Resource Availability Information:

6. Using information about the available resources (fixtures, machines, transporters, robots, tools), define the *resource vector r* as a vector with elements corresponding to all the resources involved. Determine the *resource requirements matrix F_r* for the workcell. In F_r, all elements are 0, except for 1's included in positions (i, j) if resource j is a needed for task i. In the case of no shared resources, all columns in F_r have at most a single 1.

7. Using information about which resources are needed for which tasks, determine the *resource release matrix S_r* for the workcell. In S_r, all elements are 0, except for 1's included in positions (i, j) if resource i is to be released on completion of task j. In the case of no shared resources, all rows in this matrix generally have at most a single 1. ■

In terms of these constructions, the *Rule-Based Sequencing Controller* is given by the equations:

Controller State Equation

$$x = F_v v_c + F_r r_c + F_u u_c \tag{4.15}$$

Task Start Equation

$$v_s = S_v x \tag{4.16}$$

Resource Release Equation

$$r_s = S_r x \tag{4.17}$$

Product Output Equation

$$y = S_y x \tag{4.18}$$

In these equations, subscript 'c' denotes completed tasks (v_c), or the current status of resources (r_c) or parts in (u_c). Subscript 's' denotes a command to the workcell to start tasks (v_s) or to release resources (r_s).

A graphical depiction of the rule-based controller is shown in Fig. 18, where the workcell is separated from the controller (there are some items remaining to be explained in the next subsection). The controller receives from the workcell the status of completed tasks through vector v_c, of available resources through vector r_c, and of parts in through u_c. The controller state vector x checks the conditions needed to fire the sequencing rules and hence to start the next tasks and release resources. Two vectors contain commands sent to the workcell: the next task start vector v_s indicates which tasks to start next, and r_s indicates which resources are to be released. The controller scans the rules and fires the active ones only when an *event* occurs, as is standard in DE system computer simluation (e.g. job finished, part in, etc.).

We now explain the method of computation of the controller equations. Unless properly interpreted, the equations are only a heuristic device for visualizing FMS operations; this has indeed been a major drawback of Steward's sequencing matrix and the resource requirements matrix. By the next artifice, they are turned into *rigorous mathematical equations useful for analysis and computer implementation of FMS controllers.*

272

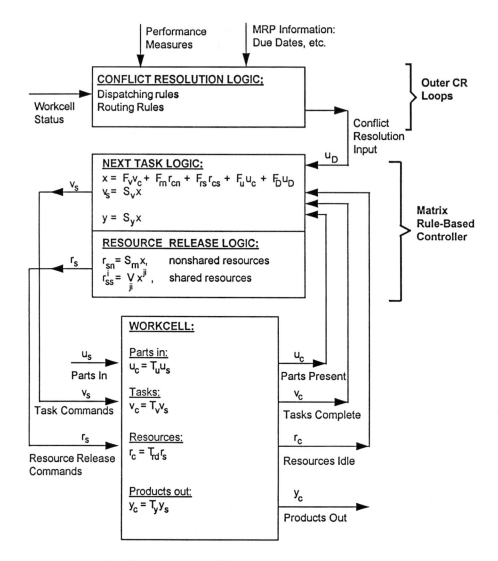

Fig. 18 Rule-Based FMS Controller

Thus, define all vectors x, v, r, u, y as expressed in *negative logic*- that is, all entries are 1 by default, and an active or high entry is set to 0. For instance, if a resource is available, its corresponding entry of r_c is set to 0. Next, the matrix operations in the controller equations are carried out in the *or/and algebra*. That is, the standard multiplication between coefficient matrices and vector elements is replaced by the logical operation 'and', and standard addition is replaced by the logical 'or'.

Example 4.1: Matrix Operations in the Or/And Algebra

The partial assembly tree of Fig. 16 has F_v given by (3.13) and F_r given by (3.14). The task vector is $v = [e\ d\ c\ b\ a]^T$ and the resource vector is $r = [F1\ F2\ R1\ R2]^T$. A formal representation of the logical controller is therefore given by (4.15), whose third row is

$$x^3 = (1 \cdot e) + (1 \cdot d) + (0 \cdot c) + (0 \cdot b) + (0 \cdot a) + (1 \cdot F1) + (0 \cdot F2) + (1 \cdot R1) + (0 \cdot R2).$$

Replacing the operations $(\cdot, +)$ by $(\wedge, \vee)$ yields

$$x^3 = (1 \wedge e) \vee (1 \wedge d) \vee (0 \wedge c) \vee (0 \wedge b) \vee (0 \wedge a) \vee (1 \wedge F1) \vee (0 \wedge F2) \vee (1 \wedge R1) \vee (0 \wedge R2),$$

whence negation and de Morgan's laws yields

$$\bar{x}^3 = (0 \vee \bar{e}) \wedge (0 \vee \bar{d}) \wedge (1 \vee \bar{c}) \wedge (1 \vee \bar{b}) \wedge (1 \vee \bar{a}) \wedge (0 \vee \bar{F}1) \wedge (1 \vee \bar{F}2) \wedge (0 \vee \bar{R}1) \wedge (1 \vee \bar{R}2)$$

or

$$\bar{x}^3 = \bar{e} \wedge \bar{d} \wedge \bar{F}1 \wedge \bar{R}1.$$

Since all vectors are represented in negative logic, this is exactly the required rule for initiating job $c = x^3$.

It is important to note that the use of negative logic and matrix or/and algebra makes the entries corresponding to 0's in F_v and F_r 'don't care' entries; for instance, the status of jobs b and a is of no concern when deciding when to fire job c. Thus, the nature of the job sequencing rules as a *partial* ordering is captured.

The operations required in the controller equations can easily be carried out using standard real-time control software on a Personal Computer, including MATLAB, MA-TRIXx, and Labview. It is noted that the coefficient matrices in (4.15) are sparse, so that real-time computations are very easy even for immense manufacturing systems. ∎

Example 4.2: Controller Design with No Shared Resources

Consider the manufacturing workcell in Fig. 13, for which the PN was drawn in Example 3.1 and the max/plus form was found in Example 3.2. The description of cell operation is given in the discussion introducing the figure. Carrying out design Algorithm 4.1 it is direct to define the task vector, input vector, output vector, and resource vector respectively as:

$$v = [M1P\ RU1\ BS\ M2P\ RU2]^T$$
$$u = PI$$
$$y = PO$$
$$r = [PA\ M1A\ M2A\ BA\ R1A\ R2A]^T,$$

where the ordering of v and r is not fixed; one possible ordering was selected in each case.

There are six sequencing rules, therefore select the controller state vector

$$x = [x^1 \ x^2 \ x^3 \ x^4 \ x^5 \ x^6]^T$$

Then, one may immediately determine the task sequencing matrix F_v, resource requirements matrix F_r, and part input matrix F_u to write the controller state equation

$$x = F_v v_c + F_r r_c + F_u u_c$$

as

$$
\begin{bmatrix} x^1 \\ x^2 \\ x^3 \\ x^4 \\ x^5 \\ x^6 \end{bmatrix}
=
\begin{bmatrix} 0 & 0 & 0 & 0 & 0 \\ 1 & 0 & 0 & 0 & 0 \\ 0 & 1 & 0 & 0 & 0 \\ 0 & 0 & 1 & 0 & 0 \\ 0 & 0 & 0 & 1 & 0 \\ 0 & 0 & 0 & 0 & 1 \end{bmatrix}
\begin{bmatrix} M1P_c \\ RU1_c \\ BS_c \\ M2P_c \\ RU2_c \end{bmatrix}
+
\begin{bmatrix} 1 & 1 & 0 & 0 & 0 & 0 \\ 0 & 0 & 0 & 0 & 1 & 0 \\ 0 & 0 & 0 & 1 & 0 & 0 \\ 0 & 0 & 1 & 0 & 0 & 0 \\ 0 & 0 & 0 & 0 & 0 & 1 \\ 0 & 0 & 0 & 0 & 0 & 0 \end{bmatrix}
\begin{bmatrix} PA_c \\ M1A_c \\ M2A_c \\ BA_c \\ R1A_c \\ R2A_c \end{bmatrix}
+
\begin{bmatrix} 1 \\ 0 \\ 0 \\ 0 \\ 0 \\ 0 \end{bmatrix} PI_c
$$

$$(4.19)$$

Even at this point, it is intriguing to compare this to the max/plus equation (3.9).

In this example, there is one rule for each job, so that matrices S_v and S_y are direct to write down, yielding the task start and output equations

$$
v_s =
\begin{bmatrix} M1P_s \\ RU1_s \\ BS_s \\ M2P_s \\ RU2_s \end{bmatrix}
=
\begin{bmatrix} 1 & 0 & 0 & 0 & 0 & 0 \\ 0 & 1 & 0 & 0 & 0 & 0 \\ 0 & 0 & 1 & 0 & 0 & 0 \\ 0 & 0 & 0 & 1 & 0 & 0 \\ 0 & 0 & 0 & 0 & 1 & 0 \end{bmatrix}
\begin{bmatrix} x^1 \\ x^2 \\ x^3 \\ x^4 \\ x^5 \\ x^6 \end{bmatrix}
= S_v x
$$

$$(4.20)$$

$$
y = PO = \begin{bmatrix} 0 & 0 & 0 & 0 & 0 & 1 \end{bmatrix}
\begin{bmatrix} x^1 \\ x^2 \\ x^3 \\ x^4 \\ x^5 \\ x^6 \end{bmatrix}
= S_y x
$$

$$(4.21)$$

Finally, taking into account the resources needed for the jobs, matrix S_r is constructed to yield the resource release equation

$$
r_s =
\begin{bmatrix} PA_s \\ M1A_s \\ M2A_s \\ BA_s \\ R1A_s \\ R2A_s \end{bmatrix}
=
\begin{bmatrix} 0 & 0 & 0 & 0 & 0 & 1 \\ 0 & 0 & 1 & 0 & 0 & 0 \\ 0 & 0 & 0 & 0 & 0 & 1 \\ 0 & 0 & 0 & 1 & 0 & 0 \\ 0 & 0 & 1 & 0 & 0 & 0 \\ 0 & 0 & 0 & 0 & 0 & 1 \end{bmatrix}
\begin{bmatrix} x^1 \\ x^2 \\ x^3 \\ x^4 \\ x^5 \\ x^6 \end{bmatrix}
= S_r x
$$

$$(4.22)$$

In Fig. 19 is shown the rule-based controller thus derived; its function is clearly separated from the workcell so that it can be implemented as code on a Personal Computer.

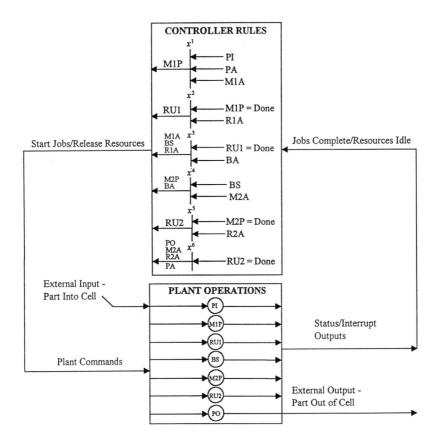

Figure 19: Rule-based FMS controller for Example 4.2

The rules are shown as PN transitions. It is interesting to compare the figure to the PN representation in Fig. 14, where the functions of the workcell and of the controller are not distinguished.

It should be clearly understood that F_v comes from the required job sequence and is usually not open to debate. On the other hand, resource matrices F_r and S_r are fixed in general structure, but some of the details could be determined differently by different production engineers. For instance, in matrix F_r moving the (3,4) entry of 1 up to position (2,4) means that *buffer availability is checked before performing move RU1*. This amounts to *kanban feedback*. In [11], [34] is given a formal method for introducing kanban feedback to avoid deadlock by modifying the structure of the resource requirements matrix F_r. ∎

4.2   Rule-based Controller for Shared Resources Case

The realistic case for manufacturing systems is when some resources are shared. In this case there are major problems in applying many of the analysis techniques in the literature. PN analysis becomes awkward in some cases, and the max/plus equations are no longer linear. We demonstrate here that the rule-based controller algorithm can easily be extended to cover shared resources. It can also cover the case of choice tasks, which includes job shops with variable part routings, as well as variable assembly sequences. The treatment of choice tasks is not given in this chapter, but appears in [14].

If there are shared resources, the FMS controller equations (4.15)-(4.18) must be extended as follows. The basic issue is that shared resources can be used for more than one job so that, on allocating these resources, one must at times make a *resource allocation choice*. This requires the addition of a *conflict resolution input* u_D (c.f. [19]). This input is selected using manufacturing *dispatching rules* [33], [21].

On the other hand, when any one of the jobs using a shared resource is completed, that shared resource is released; this requires an 'or' condition in the controller (e.g. if task a or task b is complete, then release resource c). In PN parlance, the shared-resource allocation problem corresponds to a 'fork' while the shared-resource release problem corresponds to a 'join'.

Algorithm 4.2: Controller Design with Shared Resources

1.-5. These steps are identical to those in Algorithm 4.1.

Resource Allocation Information:

6. This step is the same as in Algorithm 4.1. The only difference is that in the resource requirements matrix F_r there are some columns with more than one entry of 1; these correspond to the shared resource columns and are ordered as the last columns of F_r. Corresponding to these columns, partition matrix F_r and the resource status vector r_c as

$$F_r = [F_{rn} \quad F_{rs}]$$
$$r_c = \begin{bmatrix} r_{cn} \\ r_{cs} \end{bmatrix},$$

where a second subscript of 'n' and 's' indicates respectively nonshared resources and shared resources.

Resource Release Information:

7. The resource release matrix S_r and resource release command vector r_s are likewise conformably partitioned according to nonshared and shared resources as

$$S_r = \begin{bmatrix} S_{rn} \\ S_{rs} \end{bmatrix}$$
$$r_s = \begin{bmatrix} r_{sn} \\ r_{ss} \end{bmatrix}$$

As before, in S_r, all elements are 0, except for 1's included in positions (i, j) if resource i is to be released on completion of task j. In the case of shared resources, e.g. in matrix S_{rs}, there are multiple entries of 1 in each row, since multiple tasks release the same resource. ∎

In terms of these constructions, the rule-based controller for the case of shared resources is given by

Controller State Equation

$$x = F_v v_c + F_{rn} r_{cn} + F_{rs} r_{cs} + F_u u_c + F_D u_D \qquad (4.23)$$

Task Start Equation

$$v_s = S_v x \qquad (4.24)$$

Nonshared Resource Release Equation

$$r_{sn} = S_{rn} x \qquad (4.25)$$

Shared Resource Release Equation (nonlinear)

$$r_{ss}^i = \bigvee_{j_i} x^{j_i}, \quad \text{where } \vee \text{ denotes 'or', and } \{j_i\} \text{ is defined by } S_{rs}(i, j_i) = 1. \qquad (4.26)$$

Product Output Equation

$$y = S_y x \qquad (4.27)$$

In these equations, all vectors are in negative logic and the matrix or/and algebra is used. See Fig. 18.

In equation (4.23), the new vector u_D is a dispatching input selected to resolve conflicts between simultaneous requests for the same shared resource. Dispatching input matrix F_D is selected to match q components of u_D to each column of the shared resource requirements matrix F_{rs}, where q is the number of 1's in that column. Only one of these q entries of u_D may be high (i.e. '0') at any time, so that only one task is selected for activation based on the availability of that shared resource. The u_D is selected using dispatching rules [33], [21], [20] or using the techniques of [19].

Equation (4.26) shows how to compute the i-th component of the resource release command vector r_{ss} using 'or' operation. In this equation, the index set $\{j_i\}$ for the 'or' operation is defined to correspond to the 1 entries of matrix S_{rs} in row i. An illustration is equation (4.31) in Example 4.3, which derives from S_{rs} in (4.30). Since 'or' operations between the components of x are needed, this equation cannot be represented as a linear matrix equality in the or/and algebra. However, the expression is still easily computed, which is all we care about for FMS controller implementation.

The next example illustrates these notions.

Example 4.3: Controller Design with Shared Resources

The point of this example is to illustrate controller design for shared resources, as well as to show how to modify an existing FMS controller. It is important to see that single matrices can be modified in a very easy manner as resources change.

Consider Example 4.2, but with only one transport robot so that moves RU1 and RU2 are both performed by a single robot R. The revised resource vector is

$$r = [PA \ M1A \ M2A \ BA \ RA]^T,$$

and the resource requirements matrix becomes

$$F_r = [F_{rn} \ F_{rs}] = \begin{bmatrix} 1 & 1 & 0 & 0 & 0 \\ 0 & 0 & 0 & 0 & 1 \\ 0 & 0 & 0 & 1 & 0 \\ 0 & 0 & 1 & 0 & 0 \\ 0 & 0 & 0 & 0 & 1 \\ 0 & 0 & 0 & 0 & 0 \end{bmatrix}.$$

Note that there are two 1's in the last column, corresponding to F_{rs}, as robot R is needed for two jobs.

To resolve potential conflict arising from simultaneous requests to perform moves RU1 and RU2, one adds the outer loop term $F_D u_D$ as required by (4.23) to obtain the controller state equation

$$x = F_v v_c + F_{rn} r_{cn} + F_{rs} r_{cs} + F_u u_c + F_D u_D$$

or

$$\begin{bmatrix} x^1 \\ x^2 \\ x^3 \\ x^4 \\ x^5 \\ x^6 \end{bmatrix} = \begin{bmatrix} 0 & 0 & 0 & 0 & 0 \\ 1 & 0 & 0 & 0 & 0 \\ 0 & 1 & 0 & 0 & 0 \\ 0 & 0 & 1 & 0 & 0 \\ 0 & 0 & 0 & 1 & 0 \\ 0 & 0 & 0 & 0 & 1 \end{bmatrix} \begin{bmatrix} M1P_c \\ RU1_c \\ BS_c \\ M2P_c \\ RU2_c \end{bmatrix}$$

$$+ \begin{bmatrix} 1 & 1 & 0 & 0 & 0 \\ 0 & 0 & 0 & 0 & 1 \\ 0 & 0 & 0 & 1 & 0 \\ 0 & 0 & 1 & 0 & 0 \\ 0 & 0 & 0 & 0 & 1 \\ 0 & 0 & 0 & 0 & 0 \end{bmatrix} \begin{bmatrix} PA_c \\ M1A_c \\ M2A_c \\ BA_c \\ RA_c \end{bmatrix} + \begin{bmatrix} 0 & 0 \\ 1 & 0 \\ 0 & 0 \\ 0 & 0 \\ 0 & 1 \\ 0 & 0 \end{bmatrix} \begin{bmatrix} u_{D1} \\ u_{D2} \end{bmatrix} + \begin{bmatrix} 1 \\ 0 \\ 0 \\ 0 \\ 0 \\ 0 \end{bmatrix} PI_c \qquad (4.28)$$

where $u_D = [u_{D1} \ u_{D2}]^T$ is a conflict resolution input which is allowed to have only one of its entries high. If u_{D1} is high, then the task corresponding to x^2 is selected, if u_{D2} is high, then the task corresponding to x^5 is selected. Control input u_D is selected according to standard dispatching rules [33] that dispatch R either to move RU1 or move RU2.

In Fig. 20 is shown the modified rule-based FMS controller. There is now an outer loop corresponding to the conflict situation, which requires an external dispatching input u_D to resolve the conflict due to shared resource R.

The matrices S_v, S_y and the task start and output equations are the same as in Example 4.2.

The resource release matrices for the nonshared and shared resources are given as

$$S_{rn} = \begin{bmatrix} 0 & 0 & 0 & 0 & 0 & 1 \\ 0 & 0 & 1 & 0 & 0 & 0 \\ 0 & 0 & 0 & 0 & 0 & 1 \\ 0 & 0 & 0 & 1 & 0 & 0 \end{bmatrix} \qquad (4.29)$$

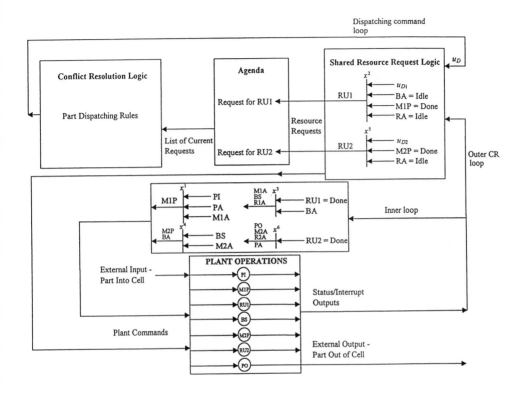

Figure 20: Rule-based FMS controller with conflict resolution loop

Figure 20: Rule-based FMS controller with conflict resolution loop

$$S_{rs} = \begin{bmatrix} 0 & 0 & 1 & 0 & 0 & 1 \end{bmatrix} \qquad (4.30)$$

The resource release commands corresponding to the nonshared resources are

$$r_{sn} = S_{rn}x$$

while those for the shared resources are manufactured according to (4.26) from S_{rs} to be

$$r_{ss} = x^3 \; .or. \; x^6. \qquad (4.31)$$

■

280

4.3 Rule-based Controller for Job Shop with Variable Part Routing

Many realistic manufacturing situations have variable part routings or a choice between multiple assembly sequences. These cases are dealt with in a straightforward manner by partitioning matrices F_v, S_v into nonchoice and choice tasks in a manner entirely similar to the partitioning of F_r, S_r in the previous subsection. The details are given in [14]. Compare also with [47], [25].

If the FMS has variable routing or multiple assembly choices, an extra conflict input is needed for choice task selection. This adds a third set of control loops in Fig. Frubafmcowicorelo. The control loops now correspond to:

1. An inner set of loops with no conflicts.
2. An intermediate set of control loops for dispatching in shared resource conflicts. This requires a dispatching conflict input u_D.
3. An outer set of control loops for routing and other task choices. This requires a routing conflict input.

It is extremely interesting to note that *dispatching and routing are dual problems* [5], [14].

In analogy to (4.26), there arise some additional nonlinear equations (e.g. requiring 'or' operations among the components of x) corresponding to the choice task start equations.

5 Properties of the Matrix FMS Controller

In this section we discuss some properties of the FMS matrix-based controller. The design algorithms given in Section 4 have shown how to write down the controller from standard IE tools such as the task sequencing matrix and resource requirements matrix. We now show that the Petri net representation of the workcell and, in decision-free cases, the max/plus equations, can be directly computed from the matrix controller. This is of extreme importance in that it effectively provides a *step-by-step design algorithm for PN and max/plus formulations for practical FMS*. This seems to be the apotheosis of design algorithms suggested in the literature such as 'top-down', 'bottom-up', etc.

It is also shown that the controller matrix formulation is of practical use in designing the structure of the controller decision loops so the controlled system is *deadlock-free*. Then, whatever dispatching rule is used, there will be no deadlock.

5.1 Petri Net Description from Controller Equations

We have seen how to construct the FMS controller equations step-by-step from standard IE tools. The next result shows that these equations are equivalent to a PN. The proof appears in [25].

Theorem 5.1
Given the FMS controller equations (4.23)-(4.27), define the *activity completion matrix*

$$F = \begin{bmatrix} F_v & F_r \end{bmatrix} = \begin{bmatrix} F_v \mid F_{rn} & F_{rs} \end{bmatrix} \tag{5.32}$$

and the *activity start matrix*

$$S = \begin{bmatrix} S_v \\ S_r \end{bmatrix} = \begin{bmatrix} S_v \\ S_{rn} \\ S_{rs} \end{bmatrix} \tag{5.33}$$

Define X as the set of elements of controller state vector x, and A (activities) as the set of elements of the task and resource vectors v and r. Then (A, X, F, S^T) is a Petri net. ∎

In fact, the theorem identifies F as the input incidence matrix and S^T as the output incidence matrix of a PN. The theorem makes it patently clear that *the PN is the closed-loop description of the workcell plus controller.* As such, all attempts to directly draw a PN for a workcell are fruitless, as they are equivalent to an attempt to draw the cell PN and design the controller all in one step.

Some more notions are required to set the stage for Section 5.3; the PN concepts are found in [36]. Thus, the PN composite change matrix is defined as M^T, where

$$M = S - F^T. \tag{5.34}$$

According to PN theory, if the initial marking vector of the net is $m(t_0)$, then the marking vector at any time $t \geq t_0$ is given by the transition equation

$$m(t) = m(t_0) + M\phi \tag{5.35}$$

where ϕ is a transition firing sequence vector. To remove ϕ from this equation, define the nullspace of M^T by a maximal matrix B such that

$$M^T B^T = (S^T - F)B^T = 0, \tag{5.36}$$

then, multiplying (5.35) on the left by B yields

$$Bm(t) = Bm(t_0). \tag{5.37}$$

This important relationship between the initial marking and the marking at any later time is known as the *loop characterizing equation.* Its solutions characterize the loops in the PN; it is nothing but a statement of the *conservative nature of the resource loops,* that is, the total number of markings in any loop is constant. In fact, the number of loops is equal to the number of rows in matrix B, and each loop is defined by one row of B.

The relation between the marking at time t and the activity vector is

$$\begin{bmatrix} v_c \\ r_c \end{bmatrix} = \ell(m) \tag{5.38}$$

where $\ell(\cdot)$ is a logical projection operator defined as

$$\ell(m_i) = \begin{cases} 0 & \text{if } m_i > 0 \\ 1 & \text{if } m_i = 0 \end{cases}.$$

That is, if a task is complete, the corresponding component of v_c is set high (denoted by 0 in negative logic): if at least one resource is available, the corresponding component of

282

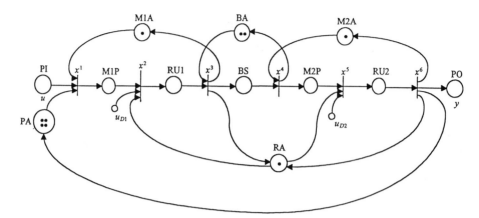

Figure 21: Petri net representation of workcell with shared resource

r_c is set high. Note that the number of resources available is captured in $m(t)$, but r_c reflects only the fact that at least one resource is available.

Example 5.1: PN and Structure from Matrix Controller Equations

Consider the controller equations from Example 4.3, where there was a shared resource R. By constructing F and S and applying the theorem, the PN in Fig. 21 is easily drawn. In this figure, initial markings have been added; this is accomplished by determining the number of resources available in the workcell.

The marking vector is defined conformably with the ordering of the columns of F as

$$m = [m_{M1P} \ m_{RU1} \ m_{BS} \ m_{M2P} \ m_{RU2} \mid m_{PA} \ m_{M1A} \ m_{M2A} \ m_{BA} \ m_{RA}]^T,$$

where m_p denotes the number of tokens in place p. Referring to Fig. 21, the initial marking vector is

$$m(t_0) = [0 \ 0 \ 0 \ 0 \ 0 \mid 4 \ 1 \ 1 \ 2 \ 1]^T$$

The activity vector corresponding to this marking is

$$
\begin{aligned}
v_c &= [1 \ 1 \ 1 \ 1 \ 1]^T \\
r_c &= [0 \ 0 \ 0 \ 0 \ 0]^T.
\end{aligned} \tag{5.39}
$$

The nullspace of M^T is defined by B^T, where

$$
B = \begin{bmatrix}
1 & 1 & 0 & 0 & 0 & 0 & 1 & 0 & 0 & 0 \\
0 & 0 & 1 & 0 & 0 & 0 & 0 & 0 & 1 & 0 \\
0 & 0 & 0 & 1 & 1 & 0 & 0 & 1 & 0 & 0 \\
0 & 1 & 0 & 0 & 1 & 0 & 0 & 0 & 0 & 1 \\
1 & 1 & 1 & 1 & 1 & 1 & 0 & 0 & 0 & 0
\end{bmatrix}. \tag{5.40}
$$

It is very interesting to examine Fig. 21 and verify that each row of B defines a resource loop.

In terms of B and $m(t_0)$, the allowed markings of the PN are given by (5.37). For instance, one solution is

$$m(t) = [0 \ 1 \ 2 \ 1 \ 0 \mid 0 \ 0 \ 0 \ 0 \ 0]^T. \tag{5.41}$$

■

5.2   Max/Plus Representation from Controller Equations

The FMS controller is concerned only with sequencing and dispatching, and does not take into account the times associated with the workcell activities. Thus, it will work for any set of activity times and covers the case of so-called 'stochastic PN'. To draw connections with the max/plus representation in Section 3.2, the activity times and resource numbers must be introduced.

Define diagonal matrices that contain the workcell activity times as follows:

$$
\begin{aligned}
T_v &= diag\{t_{v_i}\}, &&\text{task duration/setup times} \\
T_r &= diag\{t_{r_i}\}, &&\text{resource release/setup times} \\
T_u &= diag\{t_{u_i}\}, &&\text{part input times} \\
T_y &= diag\{t_{y_i}\}, &&\text{product output times.}
\end{aligned}
\tag{5.42}
$$

In these definitions, t_{v_i} is the time needed to accomplish task v_i (including setup time), t_{r_i} is the time needed to release and setup resource r_i, t_{u_i} is the time needed to move part u_i into the cell, and t_{y_i} is the time needed to move product y_i out of the cell.

In actual FMS, these times are not absolutely fixed, but may be variable or stochastic; the FMS controller of Section 4 works in all cases. However, in order to write the max/plus representation of the cell, the times must be assumed known and deterministic.

To incorporate the numbers of resources available, define the resource time/number matrix

$$
T_{rd} = diag\{t_{r_i} z^{-d_i}\} = T_r \cdot diag\{z^{-d_i}\},
\tag{5.43}
$$

where z is the unit delay operator and d_i is the number available of resource r_i.

The max/plus formulation is linear only in the case of no shared resources. Therefore, the next result shows how to obtain the max/plus formulation corresponding to the FMS controller equations (4.15)-(4.18). It is proven in [35].

Theorem 5.2
Given FMS controller equations with no shared resources (4.15)-(4.18) and the workcell time matrices T_v, T_{rd}, T_u, T_y, multiply matrices in the standard matrix algebra to obtain

$$
\begin{aligned}
x &= F_v T_v S_v x + F_r T_{rd} S_r x + F_u T_u u \\
y &= T_y S_y x.
\end{aligned}
\tag{5.44}
$$

Now, in this equation, replace all occurrences of 0 by ε and redefine all operations to be in the max/plus algebra; that is, the standard matrix operations $(\cdot, +)$ are replaced by $(+, max)$.

Then, (5.44) is the maxplus representation of the workcell-plus-controller, where x gives the time of the k-th firing of the transitions, u gives the time of k-th input of the parts, and y gives the time of k-th output of the products.

This results reveals that *the maxplus formulation is a description of the closed-loop system including both the workcell and the controller.* As such, all attempts to write it down a priori for a given workcell are fruitless, since such attempts amount to trying to describe the cell operation and design the controller all in one step.

If there are shared resources (e.g. equations (4.23)- (4.27)), then extended max/plus equations may be written that include a nonlinear equation corresponding to (4.26) [26].

Example 5.2: Max/Plus Equations from Matrix Controller Equations

For the FMS controller in Example 4.2 the task duration matrix, resource time/number matrix, and input time matrix are

$$T_v = diag\{t_{M1P}, t_{RU1}, t_{BS}, t_{M2P}, t_{RU2}\}$$
$$T_{rd} = diag\{t_{PA}z^{-4}, t_{M1A}z^{-1}, t_{M2A}z^{-1}, t_{BA}z^{-2}, t_{R1A}z^{-1}, t_{R2A}z^{-1}\}$$
$$T_u = [t_{PI}]$$

where the numbers of resources available in T_{rd} are determined from the initial markings depicted in Fig. 14. The values of the times must be determined by examining the actual workcell.

Performing now the matrix multiplication required by Theorem 5.2, followed by replacement of all occurrences of 0 by ε, yields exactly the max/plus form as given in Example 3.2. ∎

5.3 Design for Deadlock-Free Operation and Kanban Feedback

The matrix controller equations can be used for FMS design as well as analysis. To illustrate this, we show how to select the structure of the control loops so that deadlock does not occur in the FMS. Then, no matter what dispatching rules are used to select the conflict resolution input u_D, there will never be deadlock.

Any discussion on deadlock involves the structure of the rule-based controller (4.23)-(4.27), as well as information on the number of resources available, as captured in the loop characterizing equation (5.37). The relation between the controller equations and the loop equation is provided via the logical projection operator ℓ by (5.38). The next result uses controller information in the form of the activity completion matrix F defined in Theorem 5.1. It is one of several related results in [35], providing a sufficient condition for absence of deadlock that is useful in design. It is noted that necessary and sufficient conditions require more mathematical machinery, particularly some discussion on reachability.

Theorem 5.3

Let there be prescribed the FMS controller equations with shared resources (4.23)-(4.27). Compute the activity completion matrix F in (5.32) and the loop characterizing matrix B in (5.36). For the given initial resources, manufacture the initial marking vector $m(t_0)$, and determine all solutions $m(t)$ of the loop characterizing equation (5.37). Suppose that the logical projections of these admissible $m(t)$ are not contained in the nullspace of F. That is, the equation

$$F\ell(m) = 1 \tag{5.45}$$

has no solution over the matrix or/and algebra, for any m(t) satisfying (5.37). Then the FMS is deadlock-free.

Outline of Proof:

The proof is by contradiction. Deadlock is the condition where there are parts in the system but no rules can fire. If condition (5.45) holds and w is nontrivial, then according to (4.23)

$$x = F_v v_c + F_{rn} r_{cn} + F_{rs} r_{cs} = 1$$

for some nontrivial

$$w = \begin{bmatrix} v_c \\ r_{cn} \\ r_{cs} \end{bmatrix}$$

Thus, there are parts in the cell (in vector v_c), but, since x= 1, no rules are activated. This implies there is deadlock. ∎

This theorem suggests several techniques for removing deadlock, all hinging on avoiding simultaneous solutions to (5.37) and (5.45). To avoid such simultaneous solutions, one may:

1. Modify the structure of F to change its nullspace. More specifically, one modifies the structure of the resource requirements matrix F_r by moving 1's upwards in the same column; this corresponds to adding *kanban or 'lookahead'* feedback to the system.

2. Modifying the number of available initial resources $m(t_0)$ to change the solutions $m(t)$ to (5.37) [15], [46], [47]. This is generally accomplished by adding more buffer space or by restricting the number of parts in certain subsystems [34], [11].

It is not always possible to remove all deadlock from a manufacturing system. This theorem helps one do the best that is possible. Then, it makes one aware of any remaining deadlock conditions. The final avoidance of deadlock must then be accomplished by careful selection of the dispatching conflict input u_D in (4.23). Unfortunately, most existing dispatching rules are not capable of simultaneous deadlock avoidance.

Example 5.3: Deadlock-Free Design with Shared Resources

Consider the shared resource FMS of Example 4.3. It is easy to verify that the vector

$$w = [1\ 0\ 0\ 0\ 1\ 1\ 1\ 1\ 1\ 1]^T$$

satisfies condition (5.45). A marking corresponding to this activity vector is m(t) given in (5.41), for which RU1= 1, BS= 2, M2P= 1 in the PN of Fig. 21. It is esy to see that for this marking the PN is deadlocked.

Modify now the resource requirements matrix to obtain

$$F_r = [F_{rn}\ \ F_{rs}] = \begin{bmatrix} 1 & 1 & 0 & 0 & 0 \\ 0 & 0 & 0 & 1 & 1 \\ 0 & 0 & 0 & 0 & 0 \\ 0 & 0 & 1 & 0 & 0 \\ 0 & 0 & 0 & 0 & 1 \\ 0 & 0 & 0 & 0 & 0 \end{bmatrix}.$$

where the 1 in position (3,4) has been moved upwards in the same column to position (2,4). It can now be verified that there is no nontrivial solution to both (5.37) and (5.45), so that, according to Theorem 5.3, deadlock cannot occur. The removal of deadlock has been accomplished by modifying the nullspace of the resource requirements matrix F_r by changing the rule antecedent structure.

The modified resource requirements matrix F_r corresponds to the modified PN in Fig. 22. In this PN, the availability of buffer space is now checked at transition t_2 prior to beginning move RU1. This amounts to adding *kanban or 'lookahead' feedback*, and

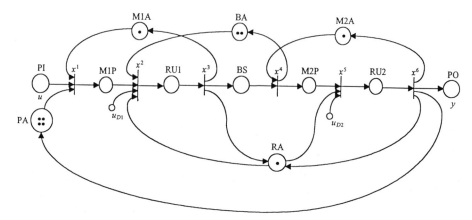

Figure 22: Petri net with shared resource and no deadlock

removes the possibility of deadlock from the FMS. Thus, kanban should be viewed as a *prefilter* for dispatching, for any dispatching rule may now be selected for input u_D to dispatch robot R to move RU1 or RU2. Whatever dispatching rule is used, deadlock cannot result.

Another approach to deadlock removal addresses the condition (5.37), modifying the intially available resources $m(t_0)$ [47]. Thus, deadlock may also be removed here by increasing the buffer space to 3 slots (e.g. set $m_{BA}(t_0) = 3$), or by limiting the number of parts in the system through reducing the number of pallets to 3 ($m_{PA}(t_0) = 3$). ∎

6 Conclusion

A new design technique has been described for rule-based controllers for flexible manufacturing systems. The controller is given in a matrix equation format where vectors are described in negative logic and mathematical computations are performed in the matrix or/and algebra. The controller is very easy to design as its coefficient matrices are nothing but Steward's task sequencing matrix [8], [44], the resource requirements matrix F_r [21], and a release release matrix, all of which are standard industrial engineering tools. The controller has a multiple loop structure, where the inner decision loops are conflict free and the outer loops have a dispatching input to resolve shared-resource conflicts. Modifying an existing FMS as resources or task objectives change is very easy, and corresponds to making changes to various individual matrices in the controller description.

The design of deadlock-free systems can be accomplished by modifying the matrix F_r, which corresponds to adding kanban feedback. Once the matrix controller has been designed, the standard Petri net formulation of the workcell-plus-controller can easily be derived. The max/plus representation is also straightforward to derive from the matrix controller.

References

[1] A. Angsana and K.M. Passino, "Distributed intelligent control of flexible manufacturing systems," *Proc. American Control Conf.*, pp. 1520-1524, June 1993.

[2] G. Balbo, G. Chiola and G. Franceschinis, "Stochastic petri net simulation for the evaluation of flexible manufacturing systems," *Proc. European Simulation MultiConf.*, pp. 5-12, Rome, June 1989.

[3] L. von Bertalanffy, *General System Theory*, New York: George Braziller, 1968.

[4] C.L.P. Chen and C. Wichman, "A CLIPS rule-based planning system for mechanical assembly," *Proc. NSF DMS Conf.*, pp. 837- 841, Atlanta, 1992.

[5] D.D. Cofer and V.K. Garg, "A timed model for the control of discrete event systems involving decisions in the Max/Plus algebra," *Proc. IEEE Conf. Decision and Control*, pp. 3363-3368, Dec. 1992.

[6] G. Cohen, D. Dubois, J.P. Quadrat, and M. Viot, "A linear- system-theoretic view of discrete-event processes and its use for performance evaluation in manufacturing," *IEEE Trans. Automat. Control*, vol. AC-30, no. 3, pp. 210-220, Mar. 1985.

[7] A.A. Desrochers, *Modeling and Control of Automated Manufacturing Systems*, IEEE Computer Society Press, 1990.

[8] S.D. Eppinger, D.E. Whitney, and R.P. Smith, "Organizing the tasks in complex design projects," *Proc. ASME Int. Conf. Design Theory and Methodology*, pp. 39-46, Sep. 1990.

[9] C.R. Glassey and S. Adiga, "Berkeley library of objects for control and simulation of manufacturing (BLOCS/M)," in *Applications of Object-Oriented Programming*, ed. L.J. Pinson and R.S. Wiener, Reading, MA: Addison-Wesley, 1990.

[10] D. Gračanin, P. Srinivasan, K. Valavanis, "Parametrized Petri nets: properties and applications to automated manufacturing systems," *Proc. IEEE Mediterranean Symp. New Directions in Control and Automation*, pp. 48- 55, June 1994.

[11] A. Gürel, O.C. Pastravanu, and F.L. Lewis, "A robust approach in deadlock-free and live FMS design," *Proc. IEEE Mediterranean Symp. New Directions in Control and Automation*, pp. 40-47, June 1994.

[12] Y.C. Ho, "Performance evaluation and perturbation analysis of discrete event dynamic systems," *IEEE Trans. Automatic Control*, vol. AC-32, pp. 563-572, 1987.

[13] Y.C. Ho, "Dynamics of discrete event systems," *Proc. IEEE. pp. 3-6*, Jan. 1989.

[14] H.-H. Huang, N. Barzin, and F.L. Lewis, "A matrix framework for manufacturing system scheduling and design," *Proc. IFAC Workshop on Intelligent Manufacturing Systems*, Vienna, June 1994.

[15] M.D. Jeng and F. DiCesare, "A synthesis method for Petri net modeling of automated manufacturing systems with shared resources," *Proc. IEEE Conf. Decision and Control*, pp. 1184- 1189, Dec. 1992.

[16] K. Jensen, "Coloured petri nets: a high level language for system design and analysis," in *Advances in Petri Nets*, ed. G. Rozenberg, Berlin: Springer-Verlag, 1990.

[17] E. Kasturia, F. DiCesare, A. Desrochers, "Real time control of multilevel manufacturing systems using colored petri nets," *Proc. IEEE Conf. Robotics Automat.*, pp. 1114-1119, 1988.

[18] M. Klein, "Supporting conflict resolution in cooperative design system," *IEEE Trans. Sys., Man, and Cybernetics*, vol.21, no. 6, pp. 1379-1390, Nov./Dec. 1991.

[19] B.H. Krogh and L.E. Holloway, "Synthesis of feedback control logic for discrete manufacturing systems,", *Automatica*, vol. 27, no. 4, pp. 641-651, July 1991.

[20] P.R. Kumar and S.P. Meyn, "Stability of queueing networks and scheduling policies," *Proc. IEEE Conf. Decision and Control*, pp. 2730-2735, Dec. 1993.

[21] A. Kusiak, "Intelligent scheduling of automated machining systems," in *Intelligent Design and Manufacturing*, ed. A. Kusiak, New York: Wiley, 1992.

[22] Y.-T. Leung and R. Suri, "Performance evaluation of discrete manufacturing systems," *IEEE Control Systems Mag.*, pp. 77-86, June 1990.

[23] F.L. Lewis, "A control system design philosophy for discrete event manufacturing systems," *Proc. Int. Symp. Implicit and Nonlinear Systems*, pp. 42-50, Arlington, TX, Dec. 1992.

[24] F.L. Lewis, H. Huang, and S. Jagannathan, "A systems approach to discrete event controller design for manufacturing systems control," *Proc. American Control Conf.*, San Francisco, pp. 1525-1531, June 1993.

[25] F.L. Lewis, O.C. Pastravanu, and H.-H. Huang, "Controller design and conflict resolution for discrete event manufacturing systems," *Proc. IEEE Conf. Decision and Control*, pp. 3288-3293, San Antonio, Dec. 1993.

[26] F.L. Lewis, H.-H. Huang, O.C. Pastravanu, and A. Gürel, "A matrix formulation for design and analysis of discrete event manufacturing systems with shared resources," *Proc. IEEE Conf. Systems, Man, and Cybernetics*, San Antonio, Oct. 1994.

[27] S.H. Lu and P.R. Kumar, "Distributed scheduling based on due dates and buffer priorities," *IEEE Trans. Automat. Control*, vol. 36, no. 12, pp. 1406-1416, Dec. 1991.

[28] P.B. Luh and D.J. Hoitomt, "Scheduling of manufacturing systems using the Lagrangian relaxation technique," *IEEE Trans. Automat. Control*, vol. 38, no. 7, July 1993.

[29] M.K. Molloy, "Performance analysis using stochastic petri nets," *IEEE Trans. Computers*, vol. C-31, no. 9, pp. 913-917, Sept. 1982.

[30] T. Murata, "Petri nets: properties, analysis and applications," *Proc. IEEE*, vol. 77, no. 4, pp. 541-580, Apr. 1989.

[31] T. Murata, N. Komoda, K. Matsumoto, and K. Haruna, "A Petri net-based controller for flexible and maintanable sequence control and its applications in factory automation," *IEEE Trans. Ind. Electronics*, vol. IE-33, no. 1, pp. 1-8, Feb. 1986.

[32] U. Negretto, "Control of manufacturing systems," in *Intelligent Design and Manufacturing*, A. Kusiak ed., New York: Wiley, 1991.

[33] S.S. Panwalker and W. Iskander, "A survey of scheduling rules," *Operations Research*, vol. 26, no. 1, pp. 45-61, Jan.-Feb. 1977.

[34] O.C. Pastravanu, A. Gürel, and F.L. Lewis, "Petri net based deadlock analysis in flow-shops with Kanban-type controllers," *Proc. ISPE/IFAC Int. Conf. on CAD/CAM, Robotics, and Factories of the Future*, Ottawa, Aug. 1994.

[35] O.C. Pastravanu, A. Gürel, H.-H. Huang, and F.L. Lewis, "Rule-based controller design algorithm for discrete event manufacturing systems," *Proc. American Control Conf.*, pp. 299-305, June 1994.

[36] J.L. Peterson, *Petri Net Theory and the Modeling of Systems*, New Jersey: Prentice-Hall, 1981.

[37] P.J. Ramadge and W.M. Wonham, "The control of discrete event systems," *Proc. IEEE*, vol. 77, pp. 81-98, 1989.

[38] S. Ramaswamy, K.P. Valavanis, P. Srinivasan, and A. Steward, "A coordination level H-EPN based error recovery model for hierarchical systems," *Proc. IEEE Mediterranean Symp. New Directions in Control and Automation*, pp. 56- 61, June 1994.

[39] P.D. Sparaggis, C.G. Cassandras, and D. Towsley, "On the duality between routing and scheduling systems with finite buffer space," *Proc. IEEE Conf. Decision and Control*, pp. 2364-2365, Dec. 1992.

[40] R.S. Sreenivas and B.H. Krogh, "On petri net models of infinite state supervisors," *IEEE Trans. Automatic Control*, vol. 37, no. 2, Feb. 1992.

[41] K. Srihari, C.R. Emerson and J.A. Cecil, "Modeling manufacturing with petri nets," *CIM Review*, pp. 15-21, Spring, 1990.

[42] K. Sycara, S. Roth, N. Sadeh, and M. Fox, "Distributed constrained heuristic search," *IEEE Trans. Sys., Man, and Cybernetics*, vol. 21, no. 6, pp. 1446-1460, Nov./Dec. 1991.

[43] S.-H. Teng and J.T. Black, "Cellular manufacturing systems modeling: the Petri net approach," *J. Manufacturing Systems*, vol. 9, no. 1, pp. 45-54, 1990.

[44] J.N. Warfield, "Binary matrices in system modeling," *IEEE Trans. Systems, Man, Cybern.*, vol. SMC-3, no. 5, pp. 441-449, Sept. 1973.

[45] J. Wolter, S. Chakrabarty, and J. Tsao, "Methods of knowledge representation for assembly planning," *Proc. NSF Design and Manuf. Sys. Conf.*, pp. 463-468, Jan. 1992.

[46] M.-C. Zhou, F. DiCesare, A.D. Desrochers, "A hybrid methodology for synthesis of Petri net models for manufacturing systems," *IEEE Trans. Robotics and Automation*, vol. 8, no. 3, pp. 350-361, Jun. 1992.

[47] M.-C. Zhou and F. DiCesare, *Petri Net Synthesis for Discrete Event Control of Manufacturing Systems*, Boston: Kluwer, 1993.

Part V

FMS Applications

Flexible Manufacturing Systems: Recent Developments
A. Raouf and M. Ben-Daya (Editors)

A Computer Integrated Robotic Flexible Welding Cell

J.E. Middle[a] and K.H. Goh[b]

[a]Department of Manufacturing Engineering, Loughborough University of Technology
Loughborough, Leicestershire, LE11 3TU, United Kingdom

[b]Rothmans International Ltd.,
Milton Keynes, United Kingdom

1 Introduction

This paper discusses the current state of development of an integrated robotic flexible welding cell (FWC) which is the result of several years of research funded by the ACME Directorate of the UK Science and Engineering Research Council. The main collaborators in the work were Cincinnati Milacron Ltd., Marchwood Engineering Laboratories of the then CEGB and the UK Welding Institute. The overall system and its information flow, excluding the robotic loading facility, is shown in Figure 1 and all elements of this are fully functioning. Briefly the system comprises off-line robot programming linked to CAD, expert systems for computer aided process planning of welding and "intelligent" process control, computer control of cell operations and provision for external communications.

2 System Hardware

The flexible welding cell is shown in Figure 2 and is based on a PLC controlled Westwood conveyor for work pallet handling, adapted to provide four work stations with precision pallet location. The Loughborough University system can handle up to six pallets although the modular conveyor is readily extended and a conveyor system of sixteen pallets is in operation at the supplier's own manufacturing facility. At the first work station loading and unloading of components is provided by a Cincinnati Milacron T3 type 566 industrial robot. The second station incorporates a laser range-finding sensor (Selcom Optcator) mounted on a three axis rectilinear manipulator. This station provides for joint inspection prior to welding and input data to one of the process control expert systems (PIKBES) [1] described in more detail later.

Welding is performed at the third station by a Fanuc S100 robot interfaced to a Gas Metal Arc welding equipment. At the fourth station inspection of the finished weld is carried out via a laser stripe sensor manipulated by a Silver Reed four axis robot.

294

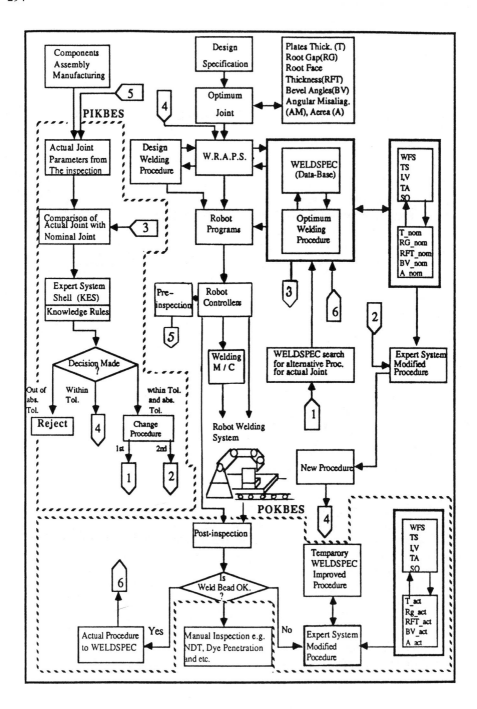

FIGURE 1 Schematic of FWC System

This provides data for another expert system (POKBES) which provides feedback for optimization of welding procedures [2].

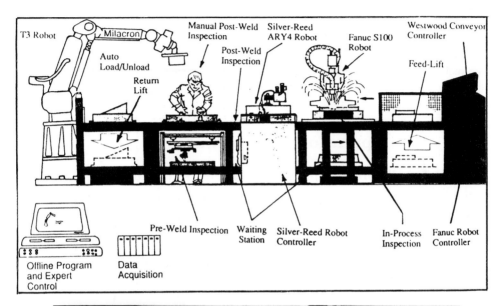

FIGURE 2 The Flexible Welding Cell.

The dedicated controllers of conveyor, robots and sensors have external computer interfaces (RS232C) and are linked via a token ring LAN to the cell supervisory controller (currently a 386 PC). Process and process equipment status and parameters are monitored continuously during welding via a Burr-Brown PCI12000 data acquisition and control board.

The off-line programming software, WRAPS (Welding Robot Adaptive Programming and Simulation) developed at Loughborough University, also provides the cell control [3]. Three further PC's in a LAN with the supervisory PC provide the computing requirements of VERSACAD [4], the CAPP expert system (ESWELDPD) [5], the welding procedure database (WELDSPEC) [6], and the process control expert systems [1,2,7].

3 System operation and softwares

3.1 WRAPS off-line programming and cell control

WRAPS was initially intended as a low cost and highly portable stand-alone package for off-line programming and simulation of robotic welding. However it has been substantially expanded for it's role as supervisory controller in the FWC and comprises four modules. It is written in the 'C' language with GSX graphics. At the time it was necessary to write special graphic application routines to execute all graphics, including 3D transformations, and utilities to design and generate icons and windows.

3.1.1 Modelling module

When manufacture of a new component is required it must first be modelled for subsequent interactive robot programming. The modelling module allows the definition of the work components as well as tooling, robots and other elements of the FWC. It is capable of modelling the required object shapes, e.g. lines, cuboid, cylinders, polyprisms, with additional shape definitions easily added by the user as required. Good graphics control for viewing the model in 3D wire frame representation is provided. When the system is in operation, a menu is displayed allowing the user easy selection of all options available. Errors encountered during use are reported instantly on the screen. The system is totally interactive, but disk files holding object data may also be read into the module for proofing or editing work. Files are stored as ASCII files so that simple geometric data files may also be created in the required format using ordinary word-processing software. Other facilities include:

- Display of co-ordinates of individuals points.

- Show distance between points.

- Creation of weld end points for use as targets in the programming module.

- Creation of pre-defined shapes such as T sections, angle sections and other standard or irregular shaped sections.

In the demonstration system at Loughborough University, component geometric data generated with the VERSACAD CAD system can be post-processed into WRAPS format saved as ASCII files and then read directly into the modelling or programming modules.

3.1.2 Programming Module

This module manipulates the models and will accept welding programming commands to produce a robot program which incorporates the tool center point (TCP) coordinates, all robot functions and all the necessary welding parameters from an expertly managed database. The TCP is defined as the tip of the welding wire electrode. The module reads the component model and any special tooling models as input files from the modelling module. The model of the particular FWC robot work station to be programmed is also retrieved and the component and tooling attached to an appropriate position on the work pallet.

During off-line robot programming work station layout can be modified to achieve optimal access for welding. In operation the model is displayed in 3D perspective with graphical representation of the welding torch attached to the robot wrist. To reduce computation time and provide smoother animation of welding movements the robot arm is usually not displayed. However joint constraint violation is evaluated during programming. Collision detection is not provided. Experience has shown that clashes are generally confined to the end effector/wrist/component region and these are readily visualized on the graphics screen. Fine tuning of the program on the actual robot is considered to be the most cost effective way of dealing with any remaining problem areas in the program.

3.2 The CAPP Function

During operation of the system the user interacts between Programming and Expert Modules to obtain the welding procedure and parameters required at each program point. The user inputs joint criteria and the system then examines automatically a hierarchy of options each of which is a rule based expert system developed in the KES expert system shell [8]. The advice given includes: Not suitable for robotic welding; Existing approved procedure available from WELDSPEC procedure database (exact match) and cost based selection where alternatives exist; Unapproved procedure available; No procedure available, input manually or invoke ESWELDPD the expert system for welding procedure design.

3.2.1 Expert Welding Procedure Designer (ESWELDPD)

This was initially developed as a stand alone CAPP facility to provide knowledge based welding procedure design and advice where no welding engineering skills are available. However it has also been fully integrated into the WRAPS Expert Module. It is menu driven and of integrated modular construction, written in the 'C' language and with embedded knowledge and rule bases and graphical routines. The main module directs the procedure design operation. Input modules allow input of joint specification and company welding capability data. This latter constraints the procedure design process to available facilities. Other modules handle process and consumable selection (constrained in the case of the FWC to those available at the robot welding station), edge preparation where this is not pre-specified and welding parameters. It outputs recommended procedures in cost based ranking. Required data from the selected procedure is read directly into WRAPS programming module and attached to weld start and end points of the off-line robot program created during programming. This includes welding arc parameters and robot welding speed. Other advice such as torch attitude can be incorporated manually through the graphical manipulation provided in the programming module.

ESWELDPD welding procedures used in programming but not empirically validated are filed in WELDSPEC as unapproved for subsequent updating following welding.

3.3 On-line Module

Communication between WRAPS and the FWC is through the On-line module. This holds the library of robot programs created off-line and post processors to format WRAPS output to that of the system robots. Communication to the robot controllers is via RS232C/current loop interface converters. This module also receives data from the cell sensory inspection systems and the monitored process parameters for use in the process control expert systems. Cell status, i.e. on/off state of equipments, availability of shielding gas, process parameters, etc. are all monitored via the Burr-Brown data acquisition board. Any abnormality in these latter inputs to WRAPS detected at any time during the running of the cell generates an output signal to the conveyor PLC. This puts it into 'hold' and consequentially, when all robot programs that are running are complete, arrests operation of the cell. Sequence control signals from the conveyor PLC are received by WRAPS via software interfaces.

4 Welding Cell operation and control

The basis of the system is the conveyor. This uses work pallets having a matrix of 8 mm holes at 25 mm centers and accurate to within 0.05 mm. These holes provide precise location of work holding fixtures on the pallets. A pallet in turn is located in its operating position by 'cup and cone' shotbolts on the underside which are actuated according to the ladder program of the conveyor PLC. Arrival of a pallet at each conveyor stage is detected by a proximity transducer which causes activation of the pallet locating shotbolts. Pallets are individually identified by a binary address using holes in the side of the pallet the pattern of which is detected by magnetic proximity transducers.

During setting up of the cell jobs are allocated to the individual pallets which are then fitted with corresponding fixtures in positions pre-determined during WRAPS programming and simulation. These job allocations are entered into WRAPS which selects the required robot programs. The supervisory computer creates in memory a circular linked list for each work station similar to the flow of pallets on the conveyor system. Slots in these lists contain the selected robot programs and other data necessary for the operations to be performed at the conveyor work stations. This includes activation of peripheral equipments and, in the case of the welding station, data from the process control expert systems. As pallets circulate in sequence on the conveyor they are identified at each work station. When the appropriate time window in the token loop LAN permits, the list corresponding to the pallet number is sent to the work station. Figure 3 shows this circular list procedure for the welding station.

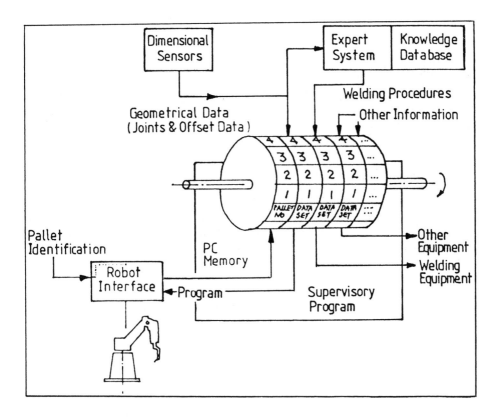

FIGURE 3 Circular Linked List for the FWC Welding Station.

A pallet arriving at the welding station is identified to the conveyor PLC for control of conveyor sequencing and shotbolt location of the pallet. The PLC sends a signal to the supervisory system which then sends the next list in the circular sequence. When the welding station's time window in token ring LAN opens data is transferred. The program is downloaded to the robot, and the robot controller communicates programmed welding parameter requirements to the welding equipments in the normal way. Other instructions will be sent from the list to switch on welding equipment, water cooler, etc. if these have been identified as in an off state via the data acquisition system. At the previous work station the joint will have been inspected and any necessary modifications to the welding procedure will have been determined by the process control expert systems. This new procedure data is communicated to the welding equipments. Finally a signal is sent to the PLC which in turn communicates with the robot control to initiate the welding robot program. During welding, welding parameters, gas flow etc. are monitored. This data is communicated to the supervisor via the data acquisition system and compared with the required values. An error signal will cause conveyor hold as described above although it is obviously possible to incorporate feedback control of most parameters. Completion of welding is identified to the PLC and conveyor sequencing then continues under it's ladder program. Similar data transfers occur at each of the work stations, the circular list sequence being matched to that of the pallets.

4.1   Loading/unloading the system

Tack welded assembled components are loaded/unloaded by the Cincinnati Milacron T3 566 robot equipped with a multi-purpose gripper. Programs are written off-line as combined unload/load routines. At the moment, on the first occasion, the robot carries out redundant unloading motions. Arrival and identification of a pallet in sequence signals the appropriate program to be sent. This is done in the same manner as described for the welding station, i.e. via the loading station's circular list, except that no additional data is needed. Fixture clamp actuation is controlled from within the robot program using the robot controller I/O facility. Failure to pick up a component is identified by complete closure of the gripper and a contact sensor. This interrupts the robot program and consequently prevents further cell operation. When the malfunction is cleared and the robot program completed cell operation can continue.

4.2   Pre-weld inspection (PIKBES)

PIKBES is a knowledge base expert system for pre-weld control of the welding procedures used by the FANUC welding robot in the cell. It is an effective tool for weld quality assurance in the cell and is described in more detail in [1,7]. Figure 4 (extracted from Figure 1) illustrates the system. Arrival of a loaded work pallet is advised to WRAPS which, through data in the station's circular list, initiates and controls scanning of the joint seam by the laser range finder (Optocator). This is mounted on the specially constructed manipulator as shown in Figure 5.

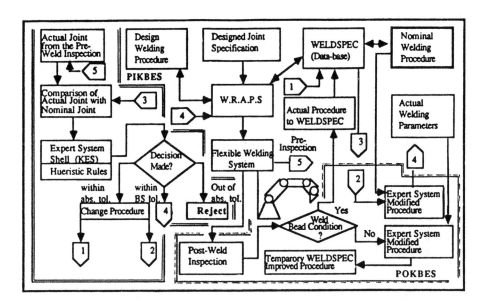

FIGURE 4 PIKBES and POKBES Process Control Expert
Systems Integrated into the FWC

FIGURE 5 Laser Rangefinding Sensor and Manipulator for PIKBES

The circular list data also controls operation of the sensor and signal processing of its analogue output. The processed sensor data are measurements of the joint profile taken at typically 5 mm intervals along the joint line. This gives dimensions of joint gap, root thickness, bevel or plate angles, plate alignment, etc. of the joint fit-up being presented for welding. This data is transferred to WRAPS expert module where it is compared with the nominal joint specification used in programming and weld procedure selection. Expert decisions are then made based on the amount of deviation detected from the nominal joint and the tolerance permitted for each joint parameter. PIKBES operates at three tolerance levels:

At level 1 the measured joint is within a tolerance at which the joint can be satisfactorily welded using the welding procedure selected during programming. PIKBES commands WRAPS to use the selected welding procedure and no new procedure data is included in the output data list to be sent to the welding station.

At level 2 the joint features are outside the tolerance at which satisfactory robotic welding is possible. WRAPS is commanded to send a signal to the conveyor PLC such that the offending work pallet will pass by the welding station and the work is not welded. The circular list in WRAPS is indexed so that it remains in sequence with the pallet flow.

At level 3, between the previous two tolerance levels, modification of the pre-selected and programmed welding procedure is required to assure satisfactory welding. The expert system, using expertly established rules, generates the required new welding procedure and passes it to the WRAPS supervisory program. It is attached to the circular list for the corresponding pallet and transferred to the welding system when the pallet is identified at the welding station. WRAPS also files this new procedure in the WELDSPEC database as an unapproved procedure.

The validity of the PIKBES knowledge base and it's effectiveness in assuring sound welds has been demonstrated on the FWC under laboratory conditions.

4.3   Post-weld inspection (POKBES)

When welding is completed on a work pallet it passes to the final work station where a laser stripe sensor, manipulated by a four axis robot, is scanned along the completed weld. Arrival of the pallet and it's identification is communicated to WRAPS signalling control data to be sent from the circular list for this inspection station. In addition to the program for the sensor manipulator, the sensor and it's peripherals are controlled. The outputs to WRAPS expert module are measurements of the finished weld profile. The post weld inspection station is shown in Figure 6.

In the expert module, POKBES, these dimensions are compared with the design specification. Conformance indicates that the welding procedure used was satisfactory for joints of the geometry measured by the pre-inspection station. If the welding procedure employed was unapproved then WRAPS will reassign it as approved in the WELDSPEC database for joints of the measured geometry.

Should the resulting weld be non-conforming the welding procedure used was unsatisfactory. The POKBES expert system will modify the procedure based on measured joint dimensions, welding procedure used and resultant weld dimensions. This modified procedure then replaces the welding procedure which had been used as an unapproved procedure in WELDSPEC.

Thus during operation of the cell existing and expert system designed new welding procedures are continually being checked and enhanced. POKBES is included in Figure 1 and 4 and described in detail in [2].

5 Concluding remarks

The system described is fully functional as a research and development facility. It's limitation as a practical FWC lies in the constrained component geometry that can currently be handled in the cell. This is primarily due to the limited manipulation provided to the inspection sensors which constrains the work to flat plate and relatively simple "cuboid" shapes.

The welding robot has joint location tactile sensing and 'through the arc' seam tracking but these are not yet incorporated in the programming capability of WRAPS. These can clearly improve the tolerance of the welding station to joint fit-up variation.

WRAPS is being developed further to give time and cost data for the various robotic operations and this is to be output to a wider management information system. The token ring LAN has a bridge for linking to such as MAP.

Finally research is on-going into the use of artificial neural networks for the process control functions of the cell.

References

[1] Ghasemshahi, M. PIKBES - Joint pre-inspection knowledge base expert system. Ph.D Thesis. Loughborough University of Technology, UK, 1991.

[2] Abdalla, K.M. Integrating inspection with a flexible Robotic Welding Cell. MSc Thesis, Loughborough University of Technology, UK, 1992.

[3] Goh, K.H. WRAPS - Welding robot adaptive programming and simulation system, Ph.D Thesis, Loughborough University of Technology, UK, 1988.

[4] VERSACAD - product of VersaCAD Corporation, 7372 Prince Drive Hundinton Beach, CA 926476, USA.

[5] Abu-Baker, N. ESWELDPD - Expert system for welding procedure design. Ph.D Thesis, Loughborough University of Technolgy, UK, 1990.

[6] WELDSPEC - Welding procedure database software. Published by the Welding Institute, UK, 1987.

[7] Middle, J.E. and Ghasemshahi, M., An Expert System Approach to the Control of Welding Procedures, International Conference on Automated Welding Systems in Manufacturing, Gateshead, UK. The Welding Institute, November, 1991.

304

[8] KES Expert system shell - Product of Software Architecture and Engineering Inc., 1600 Wilson Boulevard, Arlington, VA 22209, USA.

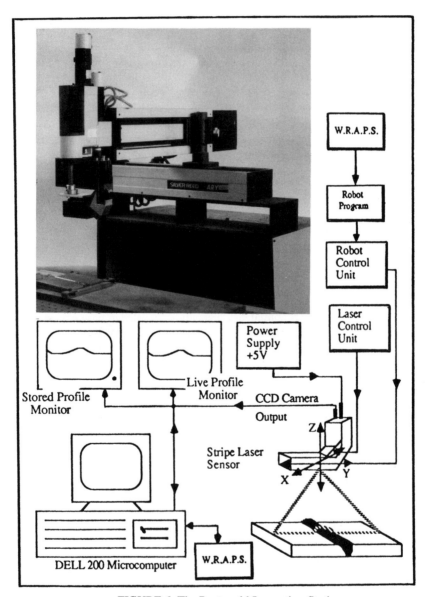

FIGURE 6 The Post-weld Inspection Station

Flexible Manufacturing Systems: Recent Developments
A. Raouf and M. Ben-Daya (Editors)
1995 Elsevier Science B.V.

Idea and Practice of Flexible Manufacturing Systems of Toyota

Atushi Masuyama

Toyota Motor Corporation, Japan

Abstract

A definition of Flexible Manufacturing System has not been necessarily clarified yet; An understanding, and an objective of its application are different in a variety of industries. In Toyota, on the basis of market-oriented production, FMS is understood as follows:

1. The ultimate objective of applying FMS is to provide a manufacturing system which may flexibly respond to changes in a market. The flexible response means to supply timely a product as demanded by a customer.

2. Therefore, we should understand FMS as an extensive system from a product design through a product distribution, but not as a system limited to a process in manufacturing.

In order to get all subsystems to work out as an organic whole, we believe that a management system should be prepared concurrently.

A general idea of FMS (FMMS), which is conducted in Toyota, is illustrated from such viewpoints as need, scope, objective, and measurement of evaluation. We also introduce various activities relating shortening the lead time in production, and in production planning and production ordering.

1 Introduction

Economic activities have long been slow on a worldwide level. It is the most critical point for manufacturers to seek, without a half and half attitude a system to produce at the least cost only as much quantity as surely to be sold so that these manufactures may win the tough business race and survive under such circumstances. It seems essential to perceive accurately the condition of the market and to supply what is demanded by the market, with a short lead time and at a low cost.

Recently, FMS has been in the spotlight. The number of companies that have introduced FMS, or have been examining doing so is increasing, which proves that many companies realize importance and difficulty of flexibly responding to changes in the market. The perception of "flexible response", however, seems to vary among companies.

In this paper, we will introduce our insight into FMS and our practical activities.

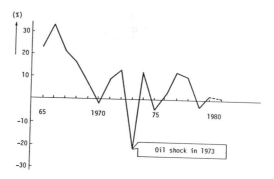

Figure 1: Annual Growth Rate - Vehicle Market in Japan

2 Why is "Flexibility" necessary?

Environment Surrounding Auto Industry

2.1 Market trend

Observing the demand of Japanese market trend of the last twenty years, there is a drastic difference between the first decade of that period and the second. The turning point was the oil crisis in 1973: the first decade can be expressed as producer oriented days in which products are usually sold out, meanwhile the second decade as consumer oriented days in which producers must be aware of over-production (Figure 1).

On the other hand, the outlook of export business is not bright either on account of the export restraint, stemmed from the trade imbalance.

Since deomestic and overseas markets become such conditions, the followings are essential to holding a dominant position in the market:

- Thoughtful expansion of product's specification so as to satisfy any customer's choice.

- Rapid resonse to a fluctuation in the quantity of the gross demand, and/or that in the quantity of various models.

- To conduct a smooth model change over.

In other words, it is required to respond to the market change with a short notice as well as to provide numerous end items (Table 1).

Table 1. Number of variants and quantity, March-May 1982.

	Variants	Quantity	Quantity/Variants
Car Line A	3,700	63,000	17
B	16,400	204,000	12
C	4,500	53,000	12
D	7,500	44,000	6
Total/Average	32,100	364,000	11

2.2 Introduction of new technology

New mechanisms of a car, new materials and new production engineering may be introduced for the sake of improvement of safety standard, fuel consumption rate, resistance against corrosion, and producibility. Such introduction requires modification in production systems and processes. It is essential that manufacturers have flexibility of accepting such modification quickly and efficiently.

3 Idea of FMS in Toyota

3.1 Objective

As mentioned above, environment (the market and technology trend) surrounding automotive manufacturing seems to become more liable to drastic changes. Hence, we believe it important that not only manufacturing cells or plants are flexible but also the entire production system should be flexibly (promptly and economically) responsive to such changes. That is, we must invent and practice a system to organize and control manufacturing cells ion which automated equipments (FMS in a narrow sense) are incorporated, which we call as Flexible Manufacturing and Management System (FMS in a broad sense). We are to make efforts to construct a flexible production system, which is based on the perception that customers decide the worth of commodities and the criterion is subject to change, depending upon circumstances.

3.2 The scope of subjects

In order to make the thorough manufacturing activities flexible, we should not consider only part of the whole process, but it is important to connect organically all relating manufacturing functions from product development planning through distribution of finished goods (Figure 2).

Focusing on production system, the following stages must carry out the functions to pursue flexibility (Figure 3).

- Production Planning
- Parts and Material Planning
- Fabrication and Assembly

308

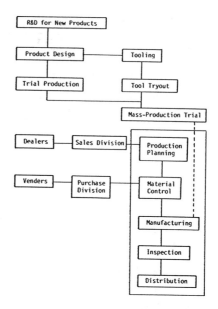

Figure 2: Structure of the Functions Relating Production.

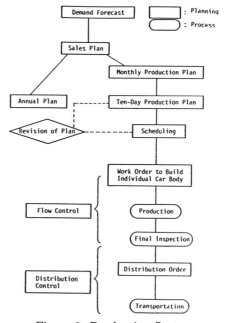

Figure 3: Production System.

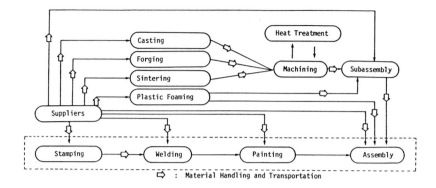

Figure 4: Outline of Production Process.

For instance, speaking about fabrication and assembly stage, making only machining shops flexible is hardly effective to improve the flexibility of the whole fabrication and assembly stage. It goes without saying that making only a machine or a machine-cell flexible is meaningless for the auto industry which consists of numerous processes to build a car (Figure 4).

When the above "flexibility concept" is extended to vendors, the ideal FMS becomes attainable. We hereafter will confine our comments to flexibility of the production system.

3.3 Measure of evaluating flexibility

We evaluate the flexibility of production system, which is based on the following view points.

3.3.1 Prompt response to a change

We appraise that to what extent the leadtime from customer's order receipt though completion of products is minimized. The lead time can be classified as follows:

A. LT_i: The lead time of processing information and communication, including demand forecast, planning production, material requirement planning, production ordering etc.

B. LT_p: The lead time of manufacturing, including machining, assembly, inspection etc.

310

C. LT_t: The lead time of transportation, including material handling and distributing finished goods.

To shorten each lead time is quite significant for flexible response to changes

$$LT \begin{cases} LT_i \\ LT_p \\ LT_t \end{cases} \to \text{Min}$$

The relation of the lead time and the accuracy of demand forecast, on which production planning is based, is generally as follows:

$$\hat{D}_t, t - LT \neq \hat{D}_t, t - LT - 1 \neq \dots \dots \neq \hat{D}_t, t - 1,$$
$$|\hat{D}_t, t - LT - D_t| > |\hat{D}_t, t - LT - 1 - D_t| \dots \dots > |\hat{D}_t, t - 1 - D_t|$$

where $D_t, t - LT$ = demand in the period t, forecasted in the period $t - LT.$, D_t = real demand in the period t.

In other words, forecast error can be minimized when one makes a demand forecast for the period t in the period as close to the period t as possible, because a trend of the market can be best incorporated to the forecast value.

3.3.2 Economical response to a change

We appraise a production system, based upon how economically it can respond to changes. "Economical response" can be defined as a condition where minimized are inventory level, and the difference between the process capacity and the actual work load level as well.

- Inventory Level $\quad \to$ Min.

- —Process Capacity - Work Load level—

What we have to keep in mind is not to let the work load level approach the process capacity, but we should consider the work load level to be determined according to the requirements of the market.

4 Actual Activities

We would like to introduce activities which we have actually developed in the areas of production planning, production ordering and production so as to attain economically shortening the lead time. Each lead time of LT_i, LT_t consists of the following three factors:

$$\left.\begin{array}{l} \text{Processing Time}(T_P) \\ \text{Setting up Time}(T_s) \\ \text{Waiting Time}(T_w) \end{array}\right\} LT$$

Activities shortening the lead times are, in short, those activities to minimize T_p and T_s, and to eliminate T_w as well.

4.1   Shortening the lead time in manufacturing process

Toyota Production System began being developed in the 1950's. Since then, we have considered that the bestg way to shorten the lead time is to practice Toyota Production System without a half and half attitude. At Toyota, the lead time from the start of body assembly to the end of the final assembly line is about one day. Practicing Just-in-time and Autoactivation in each process, which are the two basic conept of Toyota Production System, enables us to satisfy customers' requirements and to minimize in-process inventory as well.

Just-in-time is, in short, for the purpose to complete the necessary products at the necessary time through synchronization of all processes, to process only the necessary components with short lead time. To practice Just-in-time is not difficult in the case that there is only a single specification in a product. However, a manufacturer has become required by the market to provide a product with various specifications or configurations. Demand of each specification or configuration is not stable, therefore manufacturers must have mixed-model lines dealing with demand fluctuation of such a product.

It is very difficult to manage Just-in-Time in a mixed-model line. The following enable us to manage Just-in-time:

- Withdrawal by subsequent processes

- Manufacturing in a small lot

- Smoothed production

4.1.1   Withdrawal by subsequent processes

Figure 5 illustrates how production orders are given to each process in Toyota.

1. Only the first station of body assembly line receives a build-information one by one from the production control room.

2. Other processes like sub-lines or distant stations receive relevant information as process of the main line proceeds, which enables the whole production system to synchronize even if disorder of equipment or quality problem requires stpping the assembly line or re-scheduling of the production sequence. The final assembly line withdraws sub-assemblies from where those are produced as much as the final assembly line has consumed to assemble cars. The preceding shop, where those subassemblies are produced, again gives the first station of the shop replenishment order as much as withdrawn by the subsequent shop, that is, the final assembly line.

This chain-like manner connects a series of processes in a multi-stage production system. Any process other than the final assembly line never produces sub-assemblies on a forecast basis (Figure 6) Kanban is used as a tool to materialize such withdrawal by a subsequent shop.

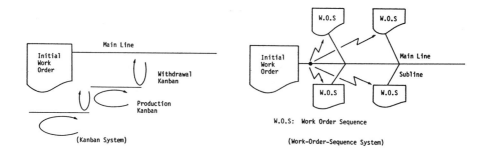

Figure 5: Information and Material Flow in Production.

Withdrawal by a subsequent shop is called as "Pull System" which has been studied in comparison with "Push System" which is a more-general production control system (Figure 6).

The characteristic of "Pull System" is to produce only the necessary material required by subsequent shops. A simulation with mathematical models has proved that fluctuations in the subsequent shops is not amplified in the preceding shops (Figure 7).

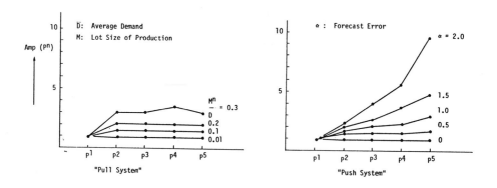

Figure 7. Comparison between "Pull System" and "Push System".

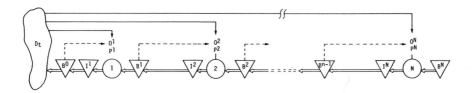

For all $n, 0^n_{t:t+L^n+1}$

$$= \hat{D}_{t:t+LT^n+\ell} + \left(\sum_{\ell=1}^{L^n} \hat{D}_{t:t+LT^{n-1}+\ell} - \sum_{\ell=1}^{L^n} P_{t-\ell:t-\ell+L^n+1} \right) - B_t^{n-1} + s^{n-1}$$

[Push System]

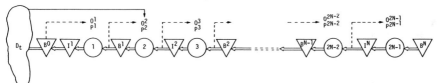

For $n = 1, 0^1_{t:t+L^1+1}$

$$= \hat{D}_{t:t+LT^1+1} + \left(\sum_{\ell=1}^{L^n} \hat{D}_{t:t+\ell} - \sum_{\ell=1}^{L^n} P_{t-\ell:t-\ell+L^1+1} \right) - B_t^0 + S^0$$

For $n > 1, 0_{t:t+L^n+1}$

$$= P_{t-L_2^{n-1}-L_1^n:t-L_1^n} + 0^n_{t-1:t+L^n} - P^n_{t-L_2^{n-1}-L_1^n-1:t-L_1^n-1}$$

[Pull System]

Figure 6. Push System and Pull System.

4.1.2 Production with a small lot size

It is essential to minimize a lot size in production at each process in order to minimize the lead time and in-process inventory as well. Reduction of setup time is crucial to minimization of a lot size. The relation of total elapsed time to setup times, the number of setups, and lot sizes is expressed as follows:

$$
\begin{aligned}
\text{T.E.T.} &= \text{total elapsed time} \\
N_i &= \text{the number of setup times of item } i \\
t_{s_i} &= \text{setup time of item } i \\
R_i &= \text{requirement of item } i \\
t_{m_i} &= \text{unit manufacturing time of item } i \\
q_i &= \text{lot size of item } i \\
N_i &= R_i/q_i \\
\text{T.E.T.} &= \sum N_i.ts_i + \sum R_i.tm_i
\end{aligned}
$$

For a given set of R_i and T.E.T., it is necessary to reduce ts_I in order to reduce q_i. Once reduction of ts_I has been achieved, more items can be manufactured for a given T.E.T.

4.1.3 Smoothed Production

When all procedures are linked by subsequent shop's withdrawals and small lot production, smoothing of product in the final assembly line with many variant's.mix is essential to minimization of capacity requirement and to elimination of excessive inventory. Consequently, gross production quantity as well as consumption rate of individual material on the final assembly line must be smoothed in terms of daily output level as well as production sequence, computing the cycle time of variants.

4.1.4 Autoactivation

Autoactivation is defined as a mechanism in which equipment or operation is designed so as to stop when abnormal conditions occur, that is, we incorporate such a device to an equipment as would sense an abnormality and stop by itself, or we give all workers a power to stop the assembly line when they find any abnormality. What we call abnormalities here are defects, delay of operation, over production, machine trouble etc. The basis of Autoactivtion concept are:

- To prevent over production

- Not to release defects to the subsequent shops

- To visualize abnormality so as to take prompt measures and to have all workers join improvement activities, which greatly contributes toward shortening the lead time in production.

4.2 Shortening the lead time in production planning and production ordering

In the case of "Push System", production planning for multi-stage production processes is generally based on $D_{t,t-\sum LT^k}$ (Figure 8).

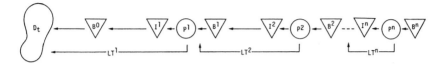

Figure 8. Production Process and Lead Time.

That is, production planning for a certain process is based on the demand forecast at the time prior to the aggregate lead time between that process and the final process. Then, the production planning is given to each process as a production order.

In the case of "Pull System", production planning for multi-stage production system is based on $D_{t,t-LT^1}$. That is, production planning is made for only the final process, and it is distributed to only the final process P^1 as a production order. Basically production orders for the other processes are given automatically through Kanban System. What enables us to conduct Kanban System is smoothing of production at the process P^1.

The difference in the lead time between "Push System" and "Pull System" results in the difference in the forecast error in the demand, as we mentioned before. The difference of the production ordering system in all processes preceding the final process results in the difference in their lead times changing production ordering when the process p^1 requires them to do so.

We, at Toyota, place great importance on especially shortening the information processing lead time in LT^1. To be more specific, it means how short the lead time from the customers' order receipt to production ordering is. In such information processing stage, we of course utilize electronic computers to develop a smoothed production planning, which, we consider, is also a sort of CAM.

5   Conclusion

The automative industry and the market has become matured, which implies that the day of a severe competition has come. In such an environment, it is necessary for the auto manufacturers to respond to various changes. We first introduced the idea of FMS in machine shops more than 20 years ago. We have gradually expanded application of such a system, with the result that Toyota and the vendors now conduct

Toyota Production System on the basis of Just-in-Time. However, we still have some problems which inders the progress of flexible production system. We would like to solve them referring to studies and researches in various fields.

References

[1] Kimura, O., and Terada, H. "Design and Analysis of Pull System, a method of multistage production control", *International Journal of Production Research*, **19**, (3) (1981), 241-253.

[2] Muramatsu, R. "The Theory and Practice of Production Management", *Journal of Japan Industrial Management Association*, **30** (4), (1980).

[3] Sugimori, Y., Kusunoki, K., Cho, F., Uchikawa, S., "Toyota Production System and Kanban System - materialization of just-in-time and respect for human system", *4th International Conference on Production Research*, (1977).